Complete Building Equipment Maintenance Desk Book

Dedication

This book is dedicated to my wife, Myrna, with love and affection for her patience, interest and cooperation

and also

to Linda, Laura, Ron, Dave, and Julia who have all put up with problems that result from such an undertaking.

Complete Building Equipment Maintenance Desk Book

Edited by
Sheldon J. Fuchs, P.E.

Prentice-Hall, Inc.   Englewood Cliffs, N.J.

Prentice-Hall International, Inc., *London*
Prentice-Hall of Australia, Pty. Ltd., *Sydney*
Prentice-Hall of Canada, Ltd.,*Toronto*
Prentice-Hall of India Private Ltd., *New Delhi*
Prentice-Hall of Japan, Inc., *Tokyo*
Prentice-Hall of Southeast Asia, Pte., Ltd., *Singapore*
Whitehall Books. Ltd., *Wellington, New Zealand*

© 1982, *by*

PRENTICE-HALL, INC.

Englewood Cliffs, N.J.

Library of Congress Cataloging in Publication Data
Main entry under title:

Complete building equipment maintenance desk book.
  Includes bibliographical references and index.
  1. Construction equipment—Maintenance and re-
pair. I. Fuchs, Sheldon J.
TH900.C6   696'.028'8   80-28190
ISBN 0-13-158808-7

Printed in the United States of America.

Acknowledgments

This author is proud to be associated with organizations that have dedicated themselves to serve their profession.

Over the years, the American Institute of Plant Engineers (AIPE) has attracted individuals who are not only professionally capable, but sensitive to the needs of fellow engineers. Several of the contributors to this manual are active participants in this society and have demonstrated their capabilities in numerous ways. Another organization that promotes high ideals and constantly updates current practices is the American Society of Heating, Refrigerating and Air Conditioning Engineers, Inc. (ASHRAE). As specialists, they have set standards and codes for the industry. Several contributors and reviewers of this book are active participants in this worthwhile society.

A third source of contributors are lecturers at the annual Hofstra University Plant Maintenance Seminar and Exhibit. Over the last ten years, thousands of plant engineers and managers have attended these conferences and heard these experts talk on various aspects of maintenance. These speakers have the capacity of combining practical know-how into a readable, hearable, understandable session.

Sheldon J. Fuchs is Plant and Facilities Manager for the Baldwin School District. His background includes Superintendent of Buildings and Grounds, Director of the Facility Maintenance Institute and adjunct professor at Hofstra University and plant maintenance engineer at Kollsman Instrument Corporation.

Mr. Fuchs holds a Bachelor of Civil Engineering and a Master's Degree in Business Administration from the City College of New York. He has continued his graduate study at New York University and Hofstra University and is a licensed professional engineer in New York State and a certified plant engineer.

His articles have appeared in many trade journals and he has been a frequent speaker at meetings and seminars of various professional engineering societies. He is a contributing author to the *Plant Engineers Manual and Guide* and the *Encyclopedia of Professional Management*. As past president of the New York and Metropolitan Manhattan chapters of the American Institute of Plant Engineers (AIPE), he has been active in various professional organizations. Mr. Fuchs is founder of the Hofstra University Annual Plant Maintenance Seminar and Exhibit.

The Editor

Contributing Authors

Robert M. Barr holds degrees from the University of North Carolina and Yale and is a member of Phi Beta Kappa. Mr. Barr has been involved with the HVAC industry for 25 years and with compressor rebuilding and parts supply for eight years. He is currently Vice-President of United Rebuilders, Inc. of Mt. Vernon, New York, and has conducted many seminars on the avoidance of compressor failure for many professional groups.

Robert Burger has a degree from Cornell and is President of Burger Associates, Inc. of Dallas, a company specializing in the rebuilding and upgrading of cooling towers. Mr. Burger has published over 30 articles that have appeared in national magazines as well as a comprehensive textbook on cooling towers entitled *Cooling Tower Technology.* He has lectured throughout the country and is a member of the Cooling Tower Institute, NAPE, NACE, ASHRAE, and other engineering societies.

William T. Cleland holds a professional engineer's license in seven states. He holds degrees from Cooper Union and New York University and is currently Commissioner of General Services for the town of Hempstead, New York. He was Executive Vice-President for Seelye, Stevenson, Value and Knecht and served as Chairman of the Town of Hempstead Housing Authority and as Vice-President of the N. Y. Association of Consulting Engineers.

Clyde H. Gordon, Sr. is an engineering consultant with Groover Engineering Co., Inc. In the maintenance services field, he has served as maintenance manager, plant engineer, and in various executive capacities including chief executive officer. As a consultant, he has designed and implemented various programs for improving cost and quality. He is a charter member of the Association of Energy Engineers.

Jack Gordon has a B.E.E. from City College of N. Y., a M.B.A. from Hofstra University and is a professional engineer in New York State. Mr. Gordon has broad experience as a utility power systems engineer, and as a facilities plant engineer. He maintains a private practice as power and energy consultant in projects ranging from major installations to electrical systems maintenance programs.

John J. Haarhaus has a B.M.E. from Villanova, and is Director of Technical Services at Lane Refrigeration. He has done early pioneering in stratosphere simulation technology and in development work in the state of the art and high altitude low temperature dew point instrumentation for the United States Air Force (U.S.A.F.). He has extensive experience in commerical and industrial refrigeration and air conditioning technology as a designer, contractor, and manufacturer's representative. Mr. Haarhaus is past president of the L. I. chapter of ASHRAE and is currently a member of the ASHRAE National Committee on Maintenance, Durability and Reliability. He is a member of the advisory boards for HVAC curriculum at State University of New York at Farmingdale, for Continuing Education in Plant Management at Hofstra University and is the chairman of Industry Committee of Refrigeration and Air Conditioning Technology at Nassau BOCES.

Frederick S. Hodgdon has a B.S. in Chemistry from M.I.T. He is the author of numerous articles and is co-author of the chapter in the ASHRAE Handbook entitled "Water and Corrosion Deposits." His specialty is in corrosion control and he is a field engineer with Lane Refrigeration Company of Island Park, New York. He is also an instructor in the Facility Maintenance Institute at Hofstra University.

Evans J. Lizardos, P.E. is a principal of the consulting engineering firm of Lizardos Engineering Associates, P.C., which specializes in process, heating, ventilating, air conditioning, electrical, plumbing, industrial energy, and related systems. He holds a B.M.E. from Brooklyn Polytechnic Institute and is a member of ASHRAE. Mr. Lizardos has been heavily involved in the conceptual planning and implementation of automatic control systems.

Charles E. Marino has a B.M.E. and a Masters in Engineering Management from City University of New York. He is a professional engineer and has a background in project engineering management in the food and pharmaceutical industries.

Robert Mayer is Vice-President of Engineering of the Tech-Lube Corp. and has developed many products, some of which have received patents. He is a graduate of Princeton and holds an M.E. from Stevens Institute and Brooklyn Polytechnic Institute. Mr. Mayer is a member of both SAE and ASLE.

Jerome Morreale has a degree in Marine Engineering from the Merchant Marine Academy at Kings Point. He is Vice-President of the MATCO Service Corporation. He was an engineering officer in the U.S. Navy and is a licensed marine engineer.

Frank L. Phillips is a senior partner of Phillips Associates of New York City. Mr. Phillips studied at the University of Chicago and Brooklyn Polytechnic Institute. He is the author of numerous technical articles and pioneered the design and development of new equipment dealing with the use of mechanical and electrical energy. Mr. Phillips received the Ecology award from the Brooklyn Engineers Club.

William Scales holds a Bachelor of Electrical Power Engineering Degree from Brooklyn Polytechnic Institute. He is the President of Scales Air Compressor

Corporation of Carle Place, New York. He has a broad background in air compressors, and is a N. Y. State licensed professional engineer. He is a member of IEEE, NSPE, and AIPE.

James H. Thompson resigned from Texas Instruments to establish J. H. Thompson Associates, Inc., a licensed consulting engineering firm specializing in security protection systems—burglary, fire and smoke, surveillance, access control, etc. He is active in organizations such as the National Fire Protection Association, National Burglar and Fire Alarm Association, Central Station Electrical Protection Association, Texas Alarm and Signal Association, and others. Mr. Thompson is a graduate of Southern Methodist University and has conducted seminars for numerous organizations on security management programs. He is a member of NSPE, and holds a professional engineer's license.

Paul D. Tomlingson is a Denver based management consultant. He is listed in the 1980 edition of *Who's Who in the West* and is a graduate of West Point. He also has an M.A. and an M.B.A. from the University of New Hampshire. His articles can be found in numerous technical journals in the profession.

W. H. Weiss has a B.S. degree in Chemical Engineering from the University of Illinois and an M.B.A. degree from Kent State University. He is a certified plant engineer and a professional engineer in the state of Ohio. Mr. Weiss presently holds the position of Corporate Maintenance Engineer for the Goodyear Tire and Rubber Company, and is the author of *Supervisor's Standard Reference Handbook*.

If You Have Maintenance Responsibility This Book Is for You

The 18 contributors to *The Complete Building Equipment Maintenance Desk Book* are outstanding examples of specialists who are well versed not only in theory but also in the how-to aspect of plant engineering and maintenance. They have pooled their talents to produce a practical information guide that will be used as a constant source of information by those who are involved in all aspects of maintenance.

This manual presents 11 chapters and two appendixes packed with ideas, checklists, guides, maintenance procedures, and concepts that will enable you to improve your operation and get the maximum from every dollar spent.

Provided are proven ideas and techniques that can *double, triple, or quadruple profits—resulting from implementing a moderate, cost effective equipment maintenance program.* Every idea, every method has been fired in the furnace of real world application. This cornucopia of practical answers offers the best thinking of a cadre of experts in the field, men who have been faced with the same problems you confront and found workable, manageable solutions. Collectively, the cost-saving, equipment-saving, manpower-saving examples from which they tap a rich rock-bed of experience, have boosted the bottom line of actual companies by hundreds of millions.

The first chapter covers a period that is overlooked by most technical references—the period spanning preparation for construction until the takeover after construction. Specific recommendations are given regarding working with the architect, how the maintenance engineer can be of most benefit during construction, and how he can develop maintenance manuals using techniques such as videotaping. The chapter contains a section on blueprint retention and preservation, with specific means of accomplishing this.

Paul D. Tomlingson in Chapter 2 explains the use of preventive maintenance (PM) in benefiting the entire maintenance program. Sample flow charts are presented that could be adapted to any operation. Steps are presented for those who desire to start a PM program. Sample forms are offered to illustrate scheduling of manpower, routes, and coordination of PM services. The use of computers in preventive maintenance scheduling is included.

Chapter 3 covers the maintenance, theory, and application of cooling towers. Robert Burger, an expert in this field, presents various checklists for the different sections of a cooling tower. In addition, recommended maintenance procedures at the beginning and at the end of the cooling seasons are included. Many suggestions for tower modifications and their advantages are presented.

Bill Scales, in Chapter 4, tells you what steps to follow to select, install, and maintain an effective compressed air system. Mr. Scales based his comprehensive checklists for air compressors and accessories on practical in-the-field experience. Chapter 4 is a must for all those concerned with the maintenance of a compressed air system whether you are an expert or not.

Chapter 5, entitled effective Motor and Automatic Control Maintenance, by Evans J. Lizardos, reveals the fundamental electrical diagram (also known as a ladder diagram) as it is used as a graphic tool to provide easy understanding of the logic of equipment control systems. The chapter leads the reader from the very simple control arrangement to the complex programmable controller applications. The author explains in a step-by-step manner the construction of elementary diagrams that frequently reveal solutions to the majority of the control problems encountered. To further clarify the maintenance procedures for motor and automatic controls, numerous illustrations are included.

In Chapter 6, three well-known experts, Jerome Morreale, Frank L. Phillips, and Robert M. Barr, cover a most complex topic. The maintenance and proper operation of air conditioning systems is divided into three sections:

- Absorption
- Centrifugal
- Reciprocating

Each of these component parts is treated in clear, precise terminology giving step-by-step procedures for operating, trouble-shooting, and maintaining the various types of equipment. These procedures are guaranteed to lessen the amount of breakdowns and to stretch the maintenance dollar. This chapter is a must for any individual involved with any aspect of air conditioning. This guide will be used not only as an occasional reference but as a day-to-day guide to efficient air conditioning maintenance procedures. To ensure easy practical help, many charts, schedules, and timetables for various overhaul procedures are included. A three-dollar, once-a-month checkup that takes 15 minutes, but can cut compressor failures by two-thirds, is revealed. A critical maintenance checklist to use after the installation of a reciprocating machine is included. This chapter also includes the cardinal rule for setting superheat, which almost every air conditioning mechanic is taught incorrectly.

In Chapter 7, Frederick Hodgdon utilizes innovative charts to show how to adjust burners in order to obtain maximum fuel burning efficiency. Proper operation and maintenance of steam and hot water boiler equipment is thoroughly discussed and simple instructive methods are given that can reduce fuel costs as much as 40 percent and significantly increase the life of the equipment. Emphasis is placed on proper maintenance of all safety equipment, which is of prime importance in the maintaining of any steam generation equipment. This practical guide written in clear language includes several logs and checklists that can be used by the operating engineer.

Jack Gordon in Chapter 8 presents an in-depth discussion of apparatus breakdown maintenance, preventative maintenance, and effective electrical subsystems operating methodology. Breakdown maintenance focuses on diagnostic techniques for failure modes, using checklists that permit the user to progress from point to point in a rational, sequential manner. System balance, optimizing of voltage, power factor, loading, emergency services, and planning for the future are discussed in precise terminology. Emphasis is placed on customer-owned substations, load and motor control centers, and rotating devices and their starter-controllers. For the benefit of the user, three major checklists for diagnostic and preventive maintenance requirements are given.

In Chapter 9, Robert Mayer demonstrates proper lubrication programs through the use of pictures and charts. Since lubrication is of vital importance in extending the life of and reducing maintenance costs for all equipment, including vehicles, heating, ventilating, air conditioning, doors, and even locks, this chapter discusses in clear, concrete terms all aspects of an effective lubrication system. The author helps you, step by step, to develop an effective lubrication scheduling procedure. Various operational gimmicks that increase efficiency such as standardization and color coding for easy recognition are revealed. Common mistakes that can lead to higher repair costs and more frequent breakdowns are discussed and analyzed.

The filter is the first line of defense in protecting heat transfer surfaces. In Chapter 10, Clyde Gordon describes the types of air filters commonly encountered in heating, ventilating, and air conditioning systems. Instructions are given on how to maintain each type of filter, as well as rules for maximizing life and economy in filter systems. Case histories for upgrading obsolete filter systems are presented to show the economics available using modern technology. Charts for standard times for testing pressure resistance across filter banks and changing filter media of various types are included. Finally, Mr. Gordon explains what to do when filtration has not been maintained, including the remedial cleaning of hot deck and cold deck coils, plenum chambers, duct work, and diffusers.

W. H. Weiss discusses pumps, bearings, fans, and belts in Chapter 11 (which are component parts of almost every piece of equipment). The breakdown of any of the above causes a complete breakdown of the operation. Therefore, it is imperative that attention be given to these items to ensure their maximum life and performance. This chapter, through the use of checklists, covers the majority of pitfalls and breakdown problems that are encountered in operational breakdowns. Emphasis is placed on a preventive maintenance approach as a method of preventing large costly breakdowns with relatively limited expenditures.

Appendix

The *Complete Building Equipment Maintenance Desk Book* features a unique addition to the 11 chapters that comprise this guide. The appendix incorporates two sections that are usually overlooked by most manuals.

In Section 1, James H. Thompson discusses security methods for safeguarding equipment. Today's society creates various problems that cannot be overlooked, for they can be a major expense to any company. How to safeguard equipment against vandalism, strikes, theft, and unauthorized access that may lead to deliberate sabotage is examined.

Section 2 by Charles E. Marino is devoted to energy-saving systems. Special attention is given to the simple inexpensive techniques to lower power requirements and reduce costs. Devices such as demand control systems and waste heat equipment are discussed. An inclusive checklist for total energy management is included.

Sheldon J. Fuchs
Editor-in-Chief

Contents

Chapter 2. PREVENTIVE MAINTENANCE: KEY TO AN EFFECTIVE, COST-SAVING MAINTENANCE PROGRAM 39

Chapter 3. COOLING TOWER TECHNOLOGY 67

Chapter 4. AIR COMPRESSORS AND THE COMPRESSED AIR SYSTEMS 103

Complete Building Equipment Maintenance Desk Book

1

The Transition Period– from Construction to Operation

Sheldon J. Fuchs, P.E.
William T. Cleland, P.E.

SCOPE OF CHAPTER

This chapter gives specific suggestions on how an engineer who becomes involved with the design or takeover of a newly constructed building or alteration can effectively preserve the design engineer's concepts. Equally important, you will learn what must be done in order to control, preserve, and organize the overwhelming instructions, manuals, guides, shop drawings, contracts, and as-built drawings that are turned over to him. Years after construction, when the architect, the engineer, and the many contractors who worked on the project are no longer available, will the maintenance manager have *meaningful documents* of his building and equipment? The documents that are referred to are not the various drawings and specifications, but rather, detailed information on how the equipment is intended to operate, its interrelationship with the building, and miscellaneous information that is not in written form.

There are many specific ways a maintenance manager can ensure he has a meaningful grasp of the entire building. This chapter deals with the importance of his participation before, during, and after the construction of both new structures and alterations.

PLANNING STAGE

The maintenance manager should have significant input into the construction plans and specifications. His expertise extends beyond maintenance. Often, he can impart data regarding the operation of the facility, information on occupational safety and health, security, energy conservation, material handling, and traffic flow. A structure should not be designed without considering the effects of vandalism. When a structure is designed, items such as civil disobedience, bomb threats, and potential strikes should be considered. If fuel oil is to be used, the size of the oil tanks should be carefully evaluated in light of the oil shortages, possible strikes, cost of oil, and other factors for which the maintenance manager has the best input. For example, the high cost of delivery should encourage the use of oversized oil storage tanks.

Unfortunately, many maintenance managers are not asked to review or comment on building construction plans and specifications. In these situations, the maintenance manager should insist on this role. He should indicate to his administrator that he can save sizable expenditures, both during the construction and in the years to come, through his expertise in the maintenance field.

Materials

In the planning stage, the maintenance manager can help to avoid serious errors that architects often make. For example, a university installed carpeting in its cafeteria. Routine policing did not remove the tremendous amount of food wedged in the fibre. The cost of constant shampooing, more than routine maintenance, and the unsightly appearance, resulted in the replacement of the carpeting with a durable inlay. Another common error often made is the use of painted surfaces in corridors and stairwells. The use of brick, stone, or various ceramic veneers can eliminate continuous painting through the years.

The use of various products such as acrylics in place of glass in high vandalism areas can result in substantial savings. Materials such as brass, large mirrored areas, or white tile require a disproportionate amount of maintenance and should be avoided.

Keep in mind the ease with which the material can be repaired and cleaned, and the cost of replacement before you approve any material.

Accessibility

Another often overlooked area of concern is accessibility. Because of an architect's oversights, economics, or a contractor's lack of concern for later consequences, the lack of accessibility can cause serious maintenance problems and affect the life of the structure. Some examples of overlooked items are:

- Metal access doors should be provided in acoustical tile ceilings with a minimum 2' x 2' dimension.
- Access doors should be provided at valves, regulators, and steam traps.
- Access to mechanical equipment rooms, fan and pump rooms, and electric rooms should be from a corridor or public space—not through an office, lounge, classroom, or janitor's closet.
- All HVAC and other operating equipment must be easily accessible for maintenance functions.

- Valves should not be hidden in ceilings nor traps sealed in walls. Panels should not have to be removed to lubricate bearings.

- Air, gas, and water lines should be installed for easy accessibility.

- Appropriate ladders should connect all roof levels to afford safe accessibility.

- Filters and light bulbs should be located so that they can be conveniently changed.

System Isolation

Make necessary provisions for adequate valves and bypasses in HVAC, plumbing, and electrical systems. Failures will occur; however, it should not be required to shut down an entire system. The installation of appropriate shut-off valves, unions and bypasses is not expensive during construction, but must be considered to isolate reasonable areas in case of a failure. All major machinery and equipment should have valves at their approach.

Clearances

In many facilities the necessary clearances are completely inadequate. One facility used a 5' aisle for all food deliveries for the feeding of seven thousand people daily. The heavy food-hand trucks continually damaged the walls, doorknobs, and other items in the corridor. The aisle should have been 8' to 10' wide. In another facility, removing the boiler tubes required the removal of a wall section. Mechanical doors and aisles should be sized adequately to insure that all equipment can be removed when necessary. Ramp angles are extremely critical. An oversized bookmobile that was purchased for a large library system had to go down a ramp and pass under an overhead door. Due to the angle and the length and height of the unit, it was unable to enter the garage.

Standardization

Standardization is imperative if operating costs are to be minimized. The advantages of having all partitions, lockers, bookshelves, HVAC equipment, controls, lighting fixtures, hardware, and other building components standardized is obvious to all maintenance managers, but not to all architects. This information can be conveyed only by the maintenance manager.

Replacement and Service

One of the most frustrating problems faced by maintenance people is having a recently installed piece of equipment fail and then finding that parts are unavailable because the model has been discontinued. In one facility, lighting fixtures were installed. The covers, when broken, were not available for replacement; either a new fixture had to be installed or the lens cover had to be especially fabricated at a very high cost. Check out newly installed equipment to make sure that parts will be available and maintenance expenses are not unreasonable. A possible solution to any questionable piece of equipment is to require an extended maintenance contract as part of the contractor's construction obligation.

Provisions for Storage and Work Space

Too often, because of oversight or other priorities, important storage space is not included. Maintenance shop areas, custodial areas, general office equipment, and miscellaneous storage areas are undersized or completely neglected. This results in significant losses because of insufficient inventories, poor organization, work inefficiencies, and numerous other losses that cannot be categorized. In equipment rooms, the installation of various overhead ceiling anchors for hoists and room for A-frames to expedite the moving of heavy equipment could pay for itself the first time it is used. These areas should be carefully planned and every effort made to include them in the facility. If the architect is given the direction to include these space requirements along with equipment requirements, he may be able to offer the owner solutions that meet all needs.

Special Requirements

This category is a catchall of practical suggestions. The following are some recommendations:

- All components, piping, and controls are to be identified by color, arrows, and tags with a key for ease of location.
- Necessary feeders for water treatment are to be included.
- Clean-out plugs in piping are to be set with approved lubricant.
- Air chambers are to be provided on hot and cold water piping at each toilet fixture.
- Entrance mats are to be recessed type.
- Hose bibs are to be provided in all toilets, machinery spaces, and 150' spacing on exterior walls.
- Plugged tees are to be used in lieu of elbows.
- Windows are to be accessible for washing and have approved anchors.
- Custodial equipment room—one room per each 10,000 square feet of floor space at each level—to have lockable storage areas; floor-type service sink with hot and cold water not less than 2'-0" above the floor and a basin curb with a minimum of 6" above the floor; a floor drain; a 3'-0" door opening outward; and recessed light fixtures at least 8' above the floor.

Omissions

Blunders or omissions are not unusual. Electrical receptacles are often forgotten, as well as appropriate pitch in roofs, sufficient numbers of roof drains, vibration dampening, and, in general, adequate provision for future expansion. These represent only some of the items to check when reviewing preliminary plans and specifications. Each facility manager can develop his own specific list according to experience. However, it must be remembered that the cost to correct items after a contract has been awarded or after construction has been completed could be considerable, unless picked up during preliminary review.

CONSTRUCTION STAGE

Too often, owners look to the architect for complete representation during all phases of construction. In some cases, an owner may request or allow his maintenance manager to check in occasionally on the construction. Yet few owners feel the necessity for a qualified employee to be assigned full-time to the construction site. Then, when problems arise, they wonder why. One important factor is the absence of the maintenance manager's formal input during the building phase. More and more, the need for the regular owner's employees to be assigned to construction along with the architect's staff is gaining acceptance. Ideally, the owner's representative should have an in-depth knowledge of mechanical and electrical building systems, a knowledge of basic construction, an ability to maintain files, and a good retention for details. His motivation would be particularly significant if he knew he would be responsible for its maintenance after completion. He would then have a vested interest in the contractor's performance, and the future maintenance of the facility.

There should be a clear understanding as to what the responsibilities of the owner's employee are. First and foremost, this representative must not interfere with the architect's clerk-of-the-works. He should not be the general liaison between the contractor and the architect. He should not involve himself with scheduling, payments, making of tests, or other routine construction matters. He must communicate any information he may have to the architect and not to the contractor.

This representative has many duties that only he can perform. He can be an involved party during the entire construction process, making sure future maintenance requirements are always considered. Often last-minute changes are made that may solve a construction problem, but will create a future maintenance nightmare. After the concrete is poured, the walls and ceilings installed, and the masonry and roofing completed, this individual should know the location of all piping such as gas, water, oil, electric, and vacuum. He should know where valves, traps, and anything else that may be permanently concealed in walls and ceilings are located. He should take his own set of job photographs and maintain his own set of as-built drawings.

During construction, the owner's representative should constantly think in terms of maintenance. He can begin to measure items such as filters, collect information on equipment maintenance, and begin to find out what mechanical manuals are available. He can check to see what spare parts are readily available, and what parts will require a long wait. In this way appropriate inventories of spare parts, including minimum and maximum levels, can be established.

The representative should make sure the needs of each department, as outlined in the planning stages, are being met. Any changes in the amount of storage space or in the layout should be brought to the attention of the prospective department head and cleared.

The owner's representative should collect, file, and catalog drawings, specifications, and important documents. He should become involved in any governmental agency, fire department regulations and various manufacturers' relationships that will affect future maintenance and operations. He should work through the architect's representative and have his suggestions, deletions, and substitutions cleared through the architect. Under no conditions should he issue orders, approve change orders or shop drawings, or perform any of the regular functions of a clerk-of-the-works.

At the conclusion of the construction, he must be completely familiar with all of the equipment and its operation, receive all manuals, drawings, and items required of the contractor, and be prepared to take over the operation of the plant. All guarantees, warranties and guaranty bonds should be received. He should have a clear concept of the maintenance program and should have determined the number of maintenance personnel required and their duties.

A master schedule should be devised for occupying the facility, making sure that all furniture, phone installations, decorations, specialized equipment, etc., are delivered and installed. During this transitional period, he should be the liaison between the new occupants and the contractors.

EQUIPMENT INVENTORY

A basic requirement in any maintenance program, whether formal or informal, is to have a complete inventory of all maintained equipment. The maintenance program is covered in another section; however, the sooner the inventory is completed, the sooner the program can be set up.

The ideal time for the inventory to be taken is during the construction phase while the equipment is being installed. The systems used for marking equipment vary from company to company. Some companies burn the number into the unit; others punch the numbers in. Other techniques are the use of metal plates, gummed plastic labels, tape, and stenciling the identifying numbers.

Excellent sources of information for the collection of inventory data and the maintenance requirements of the equipment are the manufacturers' manuals and equipment nameplates. The equipment manuals indicate recommended frequency of various kinds of maintenance. However, there are many factors that may affect the recommended frequency. Typical examples are: changing of filters in a dirty environment versus the frequency of filter changes in a clean environment, and the protective painting frequency in a corrosive atmosphere versus the frequency required in a noncorrosive atmosphere. Therefore, it is essential that the maintenance manager use the manufacturer's recommendations as a guide, while keeping in mind his specific problems and conditions.

Another major source of information is the equipment nameplates. The information should be copied as soon as possible before the nameplates become inaccessible, worn, or difficult to read. The best and most logical time for the collection of this information is when the inventory is being taken during construction.

The nameplates contain a wealth of information such as: apparatus name; manufacturer's name and city; serial number; model number; various rating information, such as nominal output and maximum ambient conditions at which output can be achieved. Some nameplates include information on installation, operation, maintenance, and replacement; others point out various cautions or warnings.

Nameplates on generators often include much significant data, such as: KW, KVA, power factors, rated armature amperes and volts, rated field currents, excitation volts, maximum temperature rises, and ambient temperatures for the armature and the field. Diesel engine-driven generator nameplates will also often include the firing order.

Nameplates can become worn, obscured, or for many reasons unreadable through the years. Therefore, when setting up the equipment files, the maintenance procedure, and the stockroom, information from the nameplates should be recorded on a permanent record file for the specific piece of equipment. If nameplate data refers to a manual, check to see if the manual is available. Many manufacturers can identify the equipment from the model and serial number on the nameplate and, in turn, can provide replacement parts lists, instructions, and rating information. If not already furnished, request the information while setting up the program. The information can be tabulated and collected easily when the equipment is installed. During this time, the manufacturers' representatives are available and any information desired can be requested.

RETENTION OF SPARE PARTS AND STORES CONTROL

An excellent opportunity is available during construction to determine what the stockroom inventory should be. Check during the installation to determine the availability of replacement items, various sources of supply, the time required to obtain the items, standard packaging practices, overhaul requirements, shelf life, space limitations, and the replacement items' values. With this information available, the replacement items for the new or remodeled structure can be incorporated into the general maintenance storeroom or whatever policy is used for the entire facility.

Keep in mind that the spare parts inventory is an insurance against prolonged shutdowns; overstocking and understocking must be avoided. Overstocking results in high carrying charges and possible losses due to obsolescence. Understocking can result in excessive downtime in case of equipment failure. It is not normally necessary to stock items for one machine when other units are in the same plant. Many manufacturers recommend minimum parts inventories for groups of machines.

The owner's representative on the construction site can be instrumental in collecting the various data and assisting in setting up the stockroom for the facility or incorporating its requirements into the overall stores operation. He can also check to find what spare parts the contractor must turn over to the owner and make sure all are received in good condition. He may recommend the owner purchase additional items that may be difficult to acquire after construction has been completed.

MAINTENANCE BY MANUALS AND VIDEO SYSTEMS

During the construction period it is wise to plan the maintenance of the facility and, after the acceptance and operation, compare the plan with the actual results. A maintenance team can be well motivated, properly trained, and inspired by working in a new facility. Without the necessary direction, many problems can develop that could have been avoided.

The ability to communicate is difficult. Too much knowledge becomes hazy. The communication of information must be more than verbal, particularly in light of the high percentage of personnel turnover in the maintenance field. For this reason the more progressive maintenance managers use devices to record (for posterity) information relative to the maintenance of a facility. The maintenance manual is that instrument. The use of videotape recordings is also being utilized. Proponents of videotaping claim the videotape can replace the maintenance manual. This is questionable, but together the maintenance manual and the videotape programs describing the maintenance and operation of the facility are an excellent combination that can maximize the best of both systems.

During construction, the owner's representative on the job should work with an experienced writer to compose the manual. The manual should not merely copy material that is readily available from any supplier. It should be practical, not theoretical, and should indicate in simple terms what is to be done, when it is to be done, and how it is to be done. The information relating to operating equipment must be readily available when needed. Basic drawings and graphs should be included. The manual must refer to the overall system as well as to the individual components. Schematic flow and/or control diagrams of each system should be indicated. The operation section may be kept separate from the maintenance section. The operation section can include a general description of the system, starting and stopping information, specific operating information, special emergency data, and various trouble-shooting information. The book can be assembled as a loose-leaf or a bound edition, and can be either a standard 8½" × 11" or 6" × 9".

It is a tremendous asset when all maintenance personnel assigned to a new or rebuilt facility can refer to a maintenance manual. Revise the initial maintenance manual constantly to ensure the incorporation of any updating or changes that occur. Comments should be solicited actively from the maintenance personnel and incorporated into the revised edition. The maintenance personnel will feel part of the revision process, and that in itself can be a substantial motivational benefit.

Much of the information conveyed by the architect, engineer, and manufacturer on the system design and intended operation can be videotaped. Specific review of the mechanical and electrical systems and material that is unique or possibly difficult to comprehend can be explained verbally and graphically on videotape. Typical questions can be asked and answered to clarify the demonstration.

The combination of the videotape lectures, a meaningful maintenance manual, and an effective supervisor who has witnessed the entire construction progress provides the ideal foundation to cost effective maintenance and construction.

RETENTION OF PLANS, SPECIFICATIONS, AND MISCELLANEOUS MATERIALS

When construction is completed, volumes of information are turned over to the owner. Some of the information is of no continuing importance as a future reference and can be discarded. However, particularly on large construction projects, hundreds of drawings, manufacturers' literature, construction specifications, catalogs, spare parts data, maintenance and operation instructions, operational test data, reports, guarantees, warranties, and other information are very valuable for future reference. This material should be incorporated into a master facility file and retained. The speed and ease with which information can be found when needed indicates the efficiency of the filing system.

Drawings in particular present many difficulties in efficient filing. Drawings may range in size from 8½″ × 11″ to 42″ × 60″. Various techniques can be used to store the original and copies. Many types of, or combinations of, cabinets and racks are designed for this purpose. Wall racks of different sizes are available. Racks house the prints either parallel or perpendicular to the wall. Various types of stands consisting of both stationary and portable models can be purchased. There are also many various models of cabinets that hold the drawings in an enclosed environment. Most of the cabinets can be locked if so desired. Plans can be stored either in a hanging position or in rolls. A cabinet of drawers in which drawings are stored horizontally is also quite common. Cabinets can be connected to drawing boards; some are designed for executive offices, and others are appropriate for drafting rooms. Cabinets can be permanent or made out of corrugated fiberboard. There are many disadvantages in using actual drawings:

- Any natural disaster such as a flood or fire can completely destroy all documentation if the files are in one location.
- The filing of and cataloging of the plans for particularly large companies may require a full-time clerk. Without this control, many drawings are borrowed and never returned.

- The continual use of drawings, particularly if no one person is responsible, results in general wear and tear of the plans. Plans will deteriorate over the years and become unreadable.

- A large area is required to store the actual plans and they usually wind up in basement areas where the dampness accelerates the deterioration.

- Good storage cabinets are expensive.

In order to overcome the cost and problems of storage and use of actual drawings, many companies are utilizing alternatives. In the last few years, the science of micrographics has been greatly improved. Micrographics is making the storage, retrieval, and dissemination of large amounts of data possible with increasing speed, efficiency, and economy. It has tremendous flexibility and can be used by both small and large companies. There are many possible variations, and the selection of the appropriate combination is up to the individual's needs and requirements.

There are many forms in which microfilm is made, stored, and used. They can be 16mm roll film, 16mm cartridge, aperture cards, 35mm film, microfiche, microfilm jackets, and micropositives. All systems have a file security, which is the ability to reproduce an inexpensive duplicate set and store it in another area so that a natural disaster does not destroy the files. Another advantage of the system is that the storage area required for the files is insignificant. A small cabinet can house thousands of copies of drawings, since microfilm reduces information 96 percent and still retains all details. Retrieval can be manual or automated. An important advantage is that once microfilmed, drawings can be reproduced at will. Therefore, assuming that the films are kept protected properly, there is no fear of loss or damage to the drawings due to misuse or age, which is so destructive to original plans.

The master microform is stored in a secure central area and duplicates are used as the active file for routine day to day use.

There are a number of different types of readers that are available to view the microfilm. Lap readers can be used for microfiche, portable readers are about the size of a typewriter, and desk and free standing units are available if larger units are desired.

Units that can read and make prints on a hard copy are available. The printer is an enlarger and can produce enlarged copies singularly or in multiple copies.

A number of options are available to the maintenance manager. For example, all work, including the microfilming of the drawings, the reading, and reproducing, can be accomplished within the company. Smaller companies can contract the microfilming out and/or the reproducing at a relatively low cost and in a short time.

If the decision is made to microfilm drawings, the entire system must be planned properly. The following questions must be considered:

1. What form will the microfilm be in?
2. How will the plans be cataloged?
3. Will the microfilming be accomplished in-house or contracted out?

The Microfilming Process

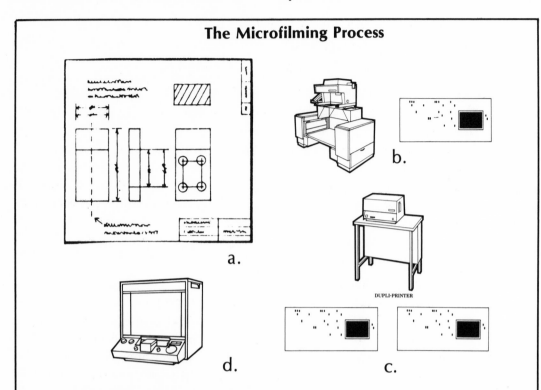

a.

b.

DUPLI-PRINTER

d.

c.

The first step (input) in the process is to reduce the original print to microfilm (Figure 1-1b) by use of an engineering camera. This camera can usually film drawings from A through E sizes reducing the drawing to a uniform size 35mm aperture card (Figure 1-1b). Some of the cameras have an internal processing that gives a finished aperture card and eliminates the processing step. The finished processed card becomes the active component of the building maintenance engineer's records control. The card can now be duplicated for any number of reasons, i.e., distribution to outlying locations, security record kept at a second area, or reference for viewing (Figure 1-1c).

The second step (output) consists of indexing or organizing the aperture cards in files for quick, easy reference. After indexing, the building maintenance engineer has an active viable alternative for records control. The aperture can be viewed for information or printed on a microfilm reader/printer onto paper for a working copy (Figure 1-1d). Any revisions to an existing document can be filmed on the camera and refiled to give an up-to-date active listing of what is taking place in any building at any point in time.

Figure 1-1. Microfilming Process:

a. Drawings to Be Reduced to Microfilm
b. Camera/Processor Produces Finished 35mm Aperture Card
c. Aperture Card Is Duplicated for Distribution
d. Duplicate Cards Are Viewed or Printed onto Paper

4. If done in-house, what camera will be used?

5. What kind of reader will be used?

6. Will the printing of the microform be done in-house or by a contractor?

7. If printing is accomplished in-house, what equipment will be used?

One recommended system would be to put drawings on aperture cards. This would have the following benefits:

A. Regardless of the original size of the drawings they are now in a one-size format.

B. Aperture cards can be color coded for separating different types of drawings. Plumbing layouts can be on buff-colored cards, electrical wiring schematics can be on yellow cards, etc.

C. Aperture cards can be written on, typed on, or keypunched for manual or sorter filing and retrieval.

D. Additions or changes can be interfiled.

E. Duplicates or duplicards can be made in seconds.

F. Several aperture cards containing pictures of drawings as large as 36" × 48" or larger can be placed in a standard number 10 envelope and mailed first class anywhere in the world for pennies.

G. Any 18" × 24" blowbacks can be made in seconds. Manufacturer's literature, descriptive material, work orders, purchase orders, correspondence, operational information, and maintenance instructions can be microfilmed onto a 16mm cartridge.

A log should be kept as to what is in each cartridge. By utilizing a seven digit number known as a microfilm access number, it is easy to keep track of the microfilmed information. The first three digits indicate the cartridge number. The second four digits indicate a particular document's location within the cartridge. Each cartridge can contain one standard file drawer of data, or approximately 3,000 pages of information.

With the new sophisticated microfilm retrieval devices, such as 3M's Page Search Reader Printer, keying in the proper location number will bring forth the specific document. An enlarged, dry copy is available in seconds if desired.

The advantages of microfilm for other records used in the maintenance department such as files, manufacturers' literature, work orders, purchase orders, and correspondence, are the same as for microfilmed drawings.

Upon the completion of a project, manufacturers' literature, descriptive material, operational information, wiring drawings, and maintenance instructions are submitted to the owner. Several copies are often submitted. It is essential that all material be filed for ready access. One copy should be maintained as a source document and never allowed to leave the office in case the other copies are lost or misplaced. The copies can be kept in files or put in loose-leaf books, although mechanics, foremen, and maintenance contractors, when using this information, generally do not preserve the copies. Like the drawings, the copies become unreadable, misplaced, or unavailable through the years. One procedure that has been particularly successful is the organizing of the material and the binding of the entire document. In this way, no one section or page could be used without taking the entire volume. The volume can be logged out to the individual like a library book. Since it is bound, no pages or drawings are lost. This procedure is inexpensive and should be considered. If the material is bound, inserts cannot be added as in a loose-leaf binder; however, a second volume can be made if new material is to be added. In one building alone, four volumes represented our library of manufacturers' information, data, maintenance recommendations, instructional sheets, parts information, bulletins, and installation information. In the rear of the volume, we inserted a library card, so that when someone borrowed the volume, the card was retained until the volume was returned.

PROJECT COMPLETION MEETING

At or near the completion of the project, a series of meetings are scheduled with the owner, the architect, the engineer, and selected suppliers or manufacturers. In addition to reviewing punch list items, basic scheduling of occupancy, and other routine matters, various design criteria are explained, along with the many operational and maintenance recommendations, comments and suggestions regarding the equipment. This series of meetings often incorporates a training program for the use of the equipment. The information that is communicated is extremely important, and in many cases it is irreplaceable and is not in any written form.

Because of the high turnover rate it is possible that the original maintenance team will no longer be employed by the firm, and a substantial amount of information that was communicated at the above-mentioned meetings is lost forever.

A solution to this dilemma which is facing more and more maintenance managers is to have the project completion meetings or any training sessions:

1. carefully structured *and*
2. videotaped.

By structuring the meetings, questions can be prepared in advance and can be answered, and general comments and recommendations can be made during the course of the meetings. The benefits of videotaping can continue for the entire life of the structure. The valuable information will be available for individuals who may need this material years later. The costs are insignificant in light of the advances in videocassette equipment.

An alternative to videocassette recording, which has built-in problems but is far less expensive, is the use of the super-eight sound movie systems now on the market. A second system to consider is the use of an instant super-eight movie system coupled with the various tape recording devices.

GUARANTEES AND WARRANTIES

The final segment of the transition phase is the period after the construction has been completed. Here, a final responsibility exists for the contractor: his responsibility regarding guarantees and warranties.

Make sure that all guarantees and warranties required by contract are submitted. These documents should be stored in a safe place. If any premature failure of items that are covered occurs, it should be made known to the responsible party. Within the first year, maintenance supervisors should be instructed to report all improperly working equipment. They should also report all structural defects, even if minor, to make sure that they are not indicative of major defects. It is a good practice to check items before their guarantee and warranty periods run out. You can keep track of guarantee or warranty expiration dates on a computer, or for smaller operations, on a supervisor's calendar. This procedure is particularly important for items that are rarely or never used. When they are needed, it sometimes becomes obvious that they have operational problems after the guarantee or warranty has expired.

All well-written specifications require a contractor to remedy any defects caused by faulty materials or workmanship, and to pay for any damages to other work resulting therefrom, which shall appear within a period of time from the date of final payment.

An owner can require a guarantee bond for longer periods on any phase of the job. In this case, the contractor must secure a surety (a third party who will ensure that the work is completed if the contractor defaults). Guarantee bonds can be required on any section of the work. For example, guarantee bonds on roofs used to be common practice. In the past few years, however, the practice of bonding roofs has been discontinued by many owners.

A warranty is frequently required with the furnishing of machinery items such as motors and pumps and in connection with the use of certain processed materials where designs are furnished by the vendor and the products are manufactured by him to meet the performance specifications. By means of a warranty the contractor certifies that the material or equipment will perform as required. The contractor is liable to a suit for damages on the grounds of a breach of warranty, if what is covered fails to perform. Surety bonds may be required to guarantee the compliance of a warranty.

Warranties can be full or limited. A product can carry more than one written warranty. For example, it can have a full warranty on part of the product and a limited warranty on the rest.

Implied warranties are rights created by state law, not by the company. All states have them.

The protection that an owner has after completion of construction is proportional to the maintenance manager's follow-up. Assuming reasonable contractual protection, if the maintenance and operation staff are diligent and all protection afforded by the contract including guarantees and warranties is pursued, it is likely that the facility under construction and its equipment will operate with minimum troubles.

MASTER PREVENTIVE MAINTENANCE SCHEDULE

LUBRICATION DESCRIPTION OF WORK	MONTHS WEEKS 1	2	3	4	5	6	7	8	9	10	11	12	13	14	15	16	17	18	19	20	21	22	23
LUBRICATION - DEPT 269																							
ROUTE 1 WEEKLY	2	2	2	2	2	2	2	2	2	2	2	2	2	2	2	2	2	2	2	2	2	2	2
ROUTE 2 WEEKLY	4	4	4	4	4	4	4	4	4	4	4	4	4	4	4	4	4	4	4	4	4	4	4
ROUTE 3 MONTHLY			2				2				2				2				2				
ROUTE 4 QUARTERLY											6												
ROUTE 5 QUARTERLY					3													3					
ROUTE 6 SEMI-ANNUALLY													4										
ROUTE 7 ANNUALLY																							
SUBTOTAL MANPOWER	6	6	8	6	9	6	8	6	6	12	8	6	10	6	8	6	9	6	8	6			
LUBRICATION - DEPT 270																							
ROUTE 1 WEEKLY	3	3	3	3	3	3	3	3	3	3	3	3	3	3	3	3	3	3	3	3	3	3	3
ROUTE 2 MONTHLY		4				4				4				4				4					
ROUTE 3 QUARTERLY						6											6						
ROUTE 4 QUARTERLY											6												
SUBTOTAL MANPOWER	3	3	7	3	9	3	7	3	3	3	7	9	3	3	7	3	9	3	7	3	3	3	
LUBRICATION - DEPT 22																							
ROUTE 1 MONTHLY	6				6				6				6				6				6		
ROUTE 2 QUARTERLY			6												6								
ROUTE 3 QUARTERLY	4												4										
SUBTOTAL MANPOWER	4	6	0	6	0	6	0	0	0	6	0	0	4	6	0	6	0	6	0	0			
TOTAL MANPOWER BY WEEKS	13	15	13	17	15	18	13	11	9	15	19	17	13	19	13	17	15	18	13				

2

Preventive Maintenance : Key to an Effective, Cost-Saving Maintenance Program

Paul D. Tomlingson

SCOPE OF CHAPTER

Preventive maintenance aims at preserving the useful life of equipment and avoiding premature failures. In addition to the routine aspects of cleaning, adjusting, lubricating, and testing, the *detection orientation* of regular equipment inspection and testing yields a vital means of improving the overall maintenance program. Early detection of impending problems results in discovery before failure and prevents a critical level of deterioration. Thus, maintenance has the opportunity to utilize this lead time to deliberately plan repair work and carry it out with more effective use of its resources. As a result, preventive maintenance inspections can become a primary means of increasing planned work, enhancing labor effectiveness, and sharply reducing emergency repairs. These conditions, in turn, reduce cost. This chapter describes the means available to the maintenance manager to achieve these objectives and the specific means of organizing and carrying out preventive maintenance functions.

PREVENTIVE MAINTENANCE DEFINED

Preventive maintenance (PM) is any action that can be taken to prolong the life of equipment and prevent premature failures.

Typically, PM includes equipment inspection, lubrication, adjustment, cleaning, nondestructive testing (predictive maintenance), and periodic maintenance—usually component replacements.

Rebuilds and overhauls are generally not classified as PM. Rather, these are actions needed when equipment has deteriorated beyond the need to conduct PM services. Rebuilding and overhauling restores equipment to an effective operating condition. Only then is PM effective.

The Purpose of Preventive Maintenance

The object of preventive maintenance is to keep equipment running more effectively and to avoid unnecessary downtime. This increases the operating time, the product output, and the life of the equipment. From a strictly maintenance view, PM has the objective of avoiding breakdown maintenance and achieving more planned work. Essentially, PM is a better way to use maintenance labor resources.

Results of a Successful PM Program

A successful preventive maintenance program can pay considerable dividends to both the maintenance department and the facility fortunate enough to have an effective PM program. To cite a few results:

- *Fewer Failures*—A timely PM inspection program usually uncovers problems before they become serious enough to cause equipment failure. As a result, routine adjustments and minor repairs take the place of failures. Breakdowns can be reduced by 50 percent.

- *More Planned Work*—The timeliness of preventive maintenance inspections usually uncovers those major jobs that require planning. Sufficient lead time allows planning to be done.

- *Fewer Emergencies*—An effective PM program has every maintenance employee on the alert for those things that cause problems. As a result, fewer problems that would cause an emergency situation escape detection. Emergency work can be reduced by half.

- *Reduced Overtime*—One of the largest contributing factors to overtime is the need to perform emergency work. A reduction in emergency work usually produces a corresponding decrease in overtime.

- *Extended Equipment Life*—PM invariably rewards its users by lasting longer and running more dependably. This usually extends equipment life by 20 percent.

- *Better Manpower Use*—A job done under emergency conditions is 15 percent more costly in labor than a similar, well-planned job. PM inspections can help to prevent emergency conditions and to increase the amount of planned work. Thus, a maintenance department with a good PM program commits more manpower to planned work. The result is more effective, productive use of this manpower.

- *Improved Equipment Operation*—Well-cared-for equipment is its own reward—it runs better.

- *Less Downtime*—Because the PM program can reduce emergency work, it follows that downtime can also be reduced. In the manufacturing environment, it is essential that an investment in scheduled downtime for PM is better than unscheduled downtime for emergency work. Once demonstrated, the PM idea becomes an accepted "better way." Then, reduced downtime can result in more production output. The "uptime" has been increased by PM.

- *Reduced Maintenance Cost*—The 15 percent extra labor cost attributable to emergency work rather than planned work must be absorbed by the maintenance department with ineffective PM. Conversely, an effective PM program can reduce these costs. Personnel tend to work more effectively with less lost motion in a well-planned job environment. This pattern soon emerges into a cost reduction trend, usually reaching about 15 percent.

PM costs will equal about 8 percent of the manhours used by a typical maintenance department and about the same percent of total maintenance costs. The overall cost of a PM program is equal to about 4 percent of total equipment value. Generally, the cost of installing a good PM program will equal about 10 percent of the total savings realized by the program.

It should be clearly understood that effective PM programs are the result of a well-organized, carefully executed effort. In short, they are hard to come by. If these benefits are to be realized, they must be earned.

PREVENTIVE MAINTENANCE INSPECTIONS

The most essential activity of the preventive maintenance program is equipment inspection. It is this portion of the PM program that generates advance information on the status of equipment. This information provides the lead time that permits a maintenance department the opportunity to plan the repairs found necessary as a result of the PM inspections.

It follows then that the preventive maintenance program should be *detection-oriented*. That is, its principal aim should be to uncover problems before they reach the crisis stage of equipment failure or breakdown. The sooner the problems are found, the greater the opportunity for planning: gathering materials, coordinating the shutdown, estimating, and allocating the manpower.

The effect of the conduct of timely PM services is illustrated in Figure 2-1.

As many PM inspection programs get underway, there is a great tendency to forget the detection-orientation. The most common action is to attempt to repair each deficiency as it is found. This tendency must be resisted because, in the long run, it undermines the PM inspection program. See Figure 2-2.

PM INSPECTIONS AND PLANNING/SCHEDULING

When PM inspections are carried out properly—on a timely basis—the lead time developed before repairs must be made provides the basis for better planning and scheduling. As a result, the amount of manpower used on scheduled work gradually increases.

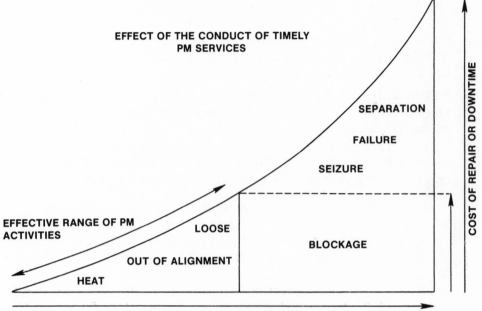

Figure 2-1.

As the deterioration of equipment increases, PM inspections, when performed on a timely basis, can uncover less serious problems before they cause damage. As a result, downtime and costs are reduced.

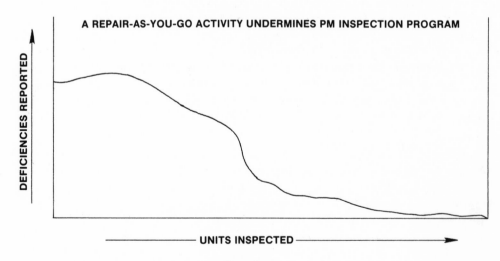

Figure 2-2.

If there were some 20 units to be inspected on a specific route, the number of deficiencies recorded would be far greater on the earlier units when repairs are made as the inspection progresses. Units at the end of the route may never get inspected because the inspector has used all of his allocated inspection time on repairs.

As the amount of scheduled work increases, so does the productive use of manpower. Essentially, as the scheduled work displaces the unscheduled and emergency work, more effective use is made of manpower. Thus, with a fixed work force more jobs are able to be performed until a point is reached when the available manpower exceeds the work that must be done. Thus, manpower savings can be generated and used elsewhere. See Figure 2-3.

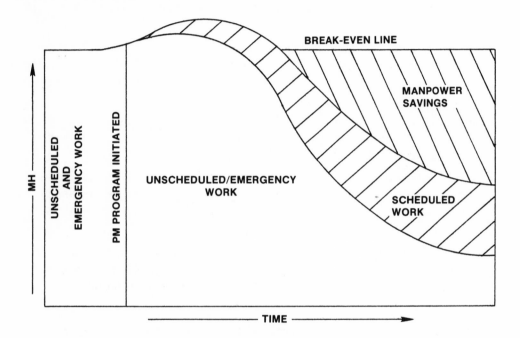

Figure 2-3.

When a PM program is initiated, there is a slight increase in the workload as PM-generated work is added to unscheduled and emergency work. Gradually, as more scheduled work is done, unscheduled and emergency work diminishes. In time, the amount of manpower for essential work falls below the previous requirement. Manpower savings result.

PM TERMINOLOGY

The following are terms used in the explanation of the development of a PM program:

- *Route*—This is the path followed by the PM worker that will ensure all equipment is reached, yet avoid needless backtracking for complete coverage.

- *Frequency or Service Interval*—This is the period of time between PM services. It can be expressed as a frequency: weekly, monthly, quarterly, etc.

- *Fixed Interval/Variable Frequency*—A fixed interval service would be carried out at the completion of a specific time period: a week, a month, or a quarter. A variable frequency service would be carried out at the end of a specific time; for example, 1,000 hours or 3,000 miles.

- *Equivalent Scheduling Day*—Many times it is easier to schedule PM services when operating hours or miles are equated to days or calendar periods (frequencies). For example, 1,000 miles could be a 30-day (monthly) service.

- *Service Inspection Time*—This is the amount of time (hours) allocated to perform the PM service.

- *Checklist*—This is the list of actions to be performed as part of the PM service.

- *Detection-Orientation*—This is the effort made to cause inspection to be the most prominent part of the PM program.

- *Visual Inspection*—This is the observation of equipment to ascertain its deficiencies.

- *Dynamic or Static Inspection*—These are inspections carried out while equipment is, respectively, in motion or at rest.

- *Drift or Chance Statistics*—These are the basic means for determining when a unit should be serviced. Drift means that the unit gradually deteriorates, and at the end of a specific period must be serviced or inspected as it is in the danger zone of failing. Chance means that there is no particular deterioration pattern, therefore, the equipment is looked at frequently enough to catch any problems occurring.

- *Workload*—This is the amount of manpower necessary to carry out the whole PM program or any of its parts, such as inspection or lubrication.

- *Nondestructive Testing*—These are testing techniques to help uncover problems using technical resources such as vibration-analysis, sonic testing, and dye testing (for cracks). Nondestructive testing is sometimes called predictive maintenance.

- *Exception Reporting*—This is PM inspection whose results lists only the problems.

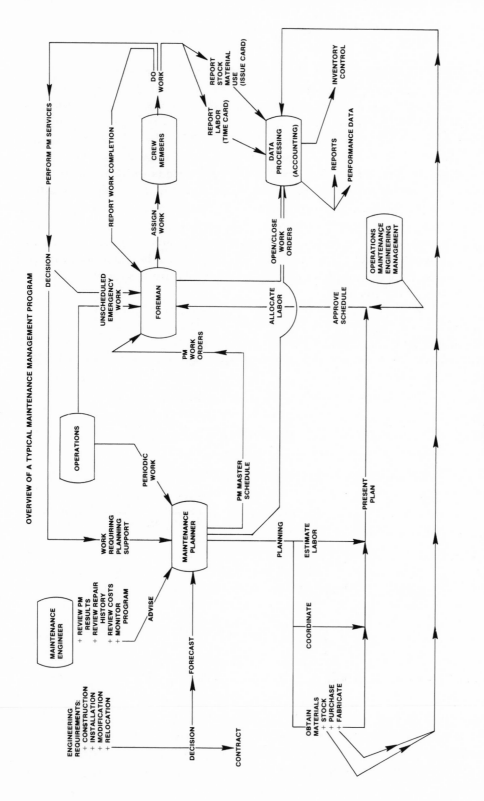

OVERVIEW OF A TYPICAL MAINTENANCE MANAGEMENT PROGRAM

Figure 2-4.

PM figures prominently within the total scope of a maintenance management network. It requires the coordinated effort of foremen and planners.

45

PM AS A PART OF THE MAINTENANCE MANAGEMENT PLAN

Just as maintenance is a key element to the production strategy, so is PM a key element to the maintenance program. In fact, to ensure that this is the case, the maintenance manager should seek strong plant management support for PM. Instead of a lukewarm endorsement such as "I'm for it," a strong endorsement is desired. A typical position might be a policy statement: "PM should take precedence over every aspect of maintenance except bona fide emergency work."

Preventive maintenance is one of several actions carried out by a maintenance department. Others include handling complaints, performing emergency or safety work, or performing periodic actions: overhauls, component replacements and rebuilds, and supporting project work.

Since these elements are fed into a maintenance department, preventive maintenance should be kept in perspective. However, because preventive maintenance inspection results usually require some planning support for major jobs, planning support is needed.

Figure 2-4 illustrates a typical Maintenance Management Network.

Within the maintenance management system is the preventive maintenance inspection control network. This network, showing the sequence of events of PM inspections, links together the actions of foremen, planners and personnel. See Figure 2-5.

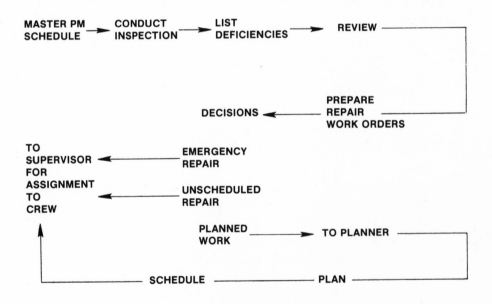

Figure 2-5.

A preventive maintenance control network traces PM from the master schedule through each event leading to completion of the work. Inspection control is illustrated.

It is absolutely necessary to ensure that the PM program fits correctly into the overall plan of the maintenance management system.

START-UP OF A PM PROGRAM

There are two broad aspects in the start-up of a PM inspection program: organizing and operating.

- *ORGANIZATIONAL* steps for starting up a PM inspection program would include actions to:

   List All Equipment to Be Inspected—The equipment to be inspected should be listed so that a basic plan can be prepared.

   Develop Routes for Fixed Equipment—Establish the route to be followed to bring the inspector to all units of equipment that must be inspected. Avoid backtracking.

   Prepare Program for Mobile Equipment—Determine the basic plan for inclusion of mobile equipment in the program.

   Establish Standard Times for Inspections—Determine how much time will be allocated to complete each service.

   Establish Intervals—Spell out how long it will be between services.

   Determine Manpower Requirements—The annual manpower requirements (workload) are equivalent to the time (hours) for each service, times the number of repetitions of that service per year.

- *OPERATING* steps for starting up a PM inspection program would include actions to:

   Prepare and Issue PM Schedules—The PM Master Schedule prescribes the week-to-week plan for the conduct of PM services. When a service is scheduled, the serviceperson uses a previously prepared checklist to conduct the service.

   Conduct Service—This is the actual conduct of prescribed services.

   Report Results—As PM services are performed, results are reported. These should be translated into work orders to correct any deficiencies uncovered.

   Monitor Repairs from PM Program—The number of repairs generated from the PM program is an indication of the thoroughness. As a result, more opportunity for scheduled work is created. Scheduled work should go up, emergencies down.

   Monitor Actual Versus Planned Times—Labor reported as a result of performing PM services should be able to be compared with the planned times. This helps to regulate the PM workload.

   Adjust Service Intervals—A lack of deficiencies when inspections occur at a certain interval is adequate license to extend inspection intervals—and the reverse is true. As this happens, review all inspection intervals and adjust them.

   Check Methods of Inspection—Determine how inspections are being done. Is there, for example, too much backtracking? Have these items checked. Industrial Engineering can help.

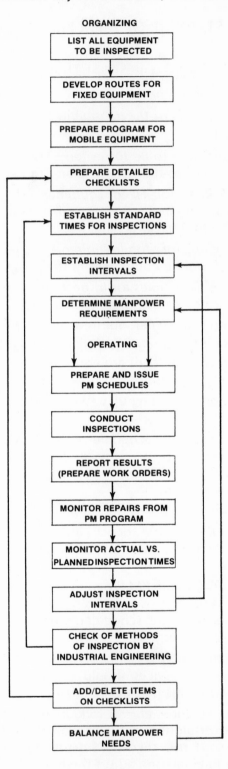

Figure 2-6.

Add/Delete Items on Checklist—Few checklists survive without change. Just as a good plan must bend to changing conditions, so must PM checklists be altered to ensure they are accurate and complete.

Balance Manpower Needs—Constantly adjust the manpower needs to satisfy PM requirements. Keep in mind that much of the PM start-up is a trial-and-error effort. Changes in manpower needs merely acknowledge this.

    See Figure 2-6.

The Details of Start-Up

Workloading is an important first step in starting up the PM program. Essentially, it is the action of finding out how much manpower is necessary to carry out the program. Unless it is done, the PM program doesn't have a chance of success. See Figure 2-7.

MASTER PREVENTIVE MAINTENANCE SCHEDULE

LUBRICATION

DESCRIPTION OF WORK (WEEKS)	1	2	3	4	5	6	7	8	9	10	11	12	13	14	15	16	17	18	19	20	21	22	23
LUBRICATION - DEPT 269																							
ROUTE 1 WEEKLY	2	2	2	2	2	2	2	2	2	2	2	2	2	2	2	2	2	2	2	2	2	2	2
ROUTE 2 WEEKLY	4	4	4	4	4	4	4	4	4	4	4	4	4	4	4	4	4	4	4	4	4	4	4
ROUTE 3 MONTHLY				2				2				2				2				2			
ROUTE 4 QUARTERLY											6												
ROUTE 5 QUARTERLY					3													3					
ROUTE 6 SEMI-ANNUALLY														4									
ROUTE 7 ANNUALLY																							
SUBTOTAL MANPOWER	6	6	6	8	6	9	6	8	6	6	12	8	6	10	6	8	6	9	6	8	6		
LUBRICATION - DEPT 270																							
ROUTE 1 WEEKLY	3	3	3	3	3	3	3	3	3	3	3	3	3	3	3	3	3	3	3	3	3	3	3
ROUTE 2 MONTHLY			4				4				4				4				4				
ROUTE 3 QUARTERLY					6												6						
ROUTE 4 QUARTERLY											6												
SUBTOTAL MANPOWER	3	3	7	3	9	3	7	3	3	3	7	9	3	3	7	3	9	3	7	3	3	3	
LUBRICATION - DEPT 22																							
ROUTE 1 MONTHLY		6				6				6				6				6					
ROUTE 2 QUARTERLY			6													6							
ROUTE 3 QUARTERLY	4												4										
SUBTOTAL MANPOWER	4	6	0	6	0	6	0	0	0	6	0	0	4	6	0	6	0	6	0	0			
TOTAL MANPOWER BY WEEKS	13	15	13	17	15	18	13	11	9	15	19	17	13	19	13	17	15	18	13				

Figure 2-7.

The estimated manpower for the conduct of preventive maintenance services must be plotted against the desired frequencies. The result provides a good picture of manpower requirements. The same procedure would be used for inspections, lubrication, or other PM services. This action balances the workload on a week-to-week basis.

Controlling PM Inspection Assignments

Master Preventive Maintenance Schedule—The Master Schedule prescribes the timing and the conduct of specific services (A, B, C, etc.) on PM inspection routes. See Figure 2-8.

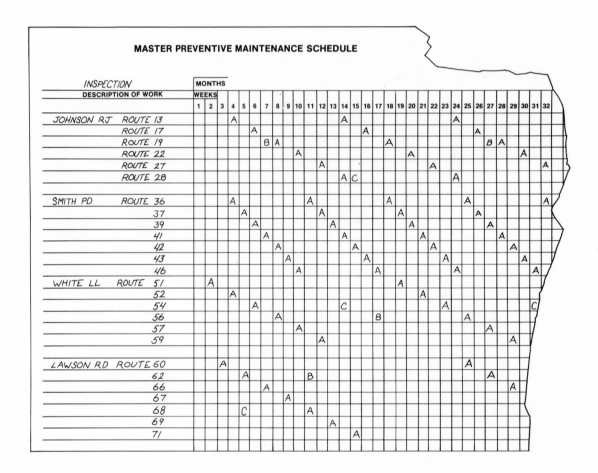

Figure 2-8.

Specific routes can be plotted against services and scheduled times. Inspectors' assignments can be made directly using this technique to simplify the foreman's job of control.

PM Inspection Checklists and Instructions

Visual inspections permit the investigative action of PM inspections to get underway before the PM checklists are written. Not only does this help to get the program underway, but it identifies inspectors who can later help prepare the checklists.

One organization asked inspectors to tape record observations as the first visual inspections were made. The results provided the basis for preparing written checklists.

Checklists often go through an evolution before any reasonably consistent (not permanent) guideline is achieved. Figure 2-9 shows how some PM checklists are started. Handwritten checklists are later typed and may end up being printed using a computer.

PM SERVICE WORKSHEET

Brown and Sharpe Grinder

PM SERVICE DESCRIPTION

ROUTE _58_   DEPT NO.   FREQUENCY _M_

GENERAL INSTRUCTIONS _Check OK, adjust or list problem. Show urgency by use of category. If work can be completed without a work order, check completed. Sign and date sheet._

EQUIPMENT DESCRIPTION	ACTION REQUIRED	OK	ADJ	PROBLEM	CAT	COMP
MAIN DRIVE	Check drive belts					
CONTROL	Check linkage & controls					
ADJUSTMENT	Check dresser mechanism					
	Check wheel spindle					
	— end —					

Figure 2-9.

This checklist is prepared on a standard worksheet. Subsequently, if desired, it can be adapted for computer use.

Figure 2-10 shows the same checklist later in its evolution as a computer-prepared checklist.

Exception Reporting means that only the problems are reported. There are many instances of equipment inspection when only the status of equipment is important. Items such as temperatures and pressures are not required. In these instances, exception reporting is simpler and quicker. Figure 2-11 shows a *PM Inspector's*

```
MWO#  91329A              PREVENTIVE MAINTENANCE WORK ORDER          ACCT  26021
DUE 07 OCT 80 DEPT 58—   UNIT 0132123045   SERVICE CODE MM
BROWN & SHARPE GRINDER                                               SCHD MM  1321


COMPLETE THE PM SERVICES DESCRIBED BELOW. CHECK OK IF THERE IS NO PROBLEM. CHECK

ADJ. IF YOU MADE AN ADJUSTMENT. WRITE IN PROBLEM WHERE YOU FIND A PROBLEM.

OPPOSITE THE PROBLEM WRITE A 4 UNDER CATEGORY IF THE REPAIR CAN BE MADE WITHIN

ONE WEEK. WRITE A 5 UNDER CATEGORY IF THE REPAIR MUST BE MADE WITHIN ONE DAY.

RETURN THIS CARD TO YOUR SUPERVISOR. REPORT MWO #, TIME AND ACCOUNT # ON YOUR

WORK RECORD CARD.

SIGNATURE  RL Smith          DATE 10/17/80        SHIFT    1 ✗ 3
```

```
MWO#  91329A              PREVENTIVE MAINTENANCE WORK ORDER          ACCT 26021
DUE 07 OCT 80 DEPT 58—   UNIT 0132123045   SERVICE CODE MM
BROWN & SHARPE GRINDER                                               SCHD MM  1321

COMPONENT ACTION REQUIRED              OK  ADJ  PROBLEM                     CAT

   1 DRIVE BELTS
   2 OPERATING LINKAGE & CONTROL
   3 DRESSER MECHANISM
   4 WHEEL SPINDLE BEARINGS

                           —END OF SERVICE LIST—
```

Figure 2-10.

The PM checklist in this illustration is printed by the computer on punch card-sized forms. They appear in the form of a work order that is automatically opened when the PM work order is printed.

Worksheet in which the inspector, working from a printed checklist, has reported only the problems he has found. This places his foreman in the position of having to take action directly on every item. Thus, a more positive action can result.

Getting Help in Starting the PM Program

Don't forget to involve the users of the equipment. These personnel have a number of key activities to perform in making the PM program successful.

- *The Scope of PM Work*—How much PM work can be done and how can nonmaintenance help sustain it by making equipment available? Perhaps one of the most prevalent causes of the failure of PM programs is the absence of operator participation at the outset. Few PM programs ever

PM INSPECTOR'S WORKSHEET

*INSTRUCTIONS: Complete log sheet. Check results against existing cards. Line out duplications.
Prepare cards on all others. Follow standard scheduling procedure.*

ROUTE ___270-4___   WEEK NO. __32__

EQUIPMENT/ LOCATION	COMPONENT	DESCRIPTION OF WORK NEEDED	IN WKS
1471		vari-drive base cracking – weld	2
1202		pillow-block brg. out of alignment	1
7114		clean spilled oil / under panel	1
3,618		replace spring return on skip la...	1
		...use sight	

Figure 2-11.

The PM inspector lists only those problems he finds. Thus, action is required on each item listed.

succeed as a unilateral maintenance effort. A joint effort is essential to identify PM services required, their frequency—and their potential benefit.

- *Frequency*—How often will services be performed? Can users make units available? If not, what alternate scheduling arrangements can be made?

- *Availability of Equipment for Service*—When visual inspections are made there is little problem in making equipment available. Unfortunately, most PM service requires that the equipment be "down"—often locked-out. This time is downtime also, and it must be treated as such—planned and prepared for. The PM program must be coordinated to ensure scheduling of PM services.

- *Downtime Allocation for PM*—Operating time can be attained only by equipment that operates. A particular level of production, for example, must be accompanied by a specific period of utilized time of equipment. Therefore, equipment users must be as aware of scheduled PM downtime as they are of scheduled running time.

- *Actions of Operators*—Operators know their equipment well. They are aware of its many quirks. They are prepared to tell PM inspectors of many suspected problems. Thus, they are an excellent source of information on equipment condition. They can also check lube or oil levels, test hydraulic controls, or verify safety devices. Many times they can carry out daily lube services when it is difficult for regular lubricators to reach the equipment on a timely basis. The point is—these operators are an excellent source of supplemental PM support, and should be used.

- *Review of Deficiency Lists*—When serious problems are found during a PM inspection, the equipment operator or his supervisor has a right to know so that he can plan around the subsequent repair. He may have to curtail planned use of the equipment to accommodate the deficiency.

- *Decisions on Repairs/Scheduling*—Once major deficiencies are uncovered, repairs must be made. These repairs also need downtime—and they must be coordinated with equipment users.

PM Inspector

The PM inspector is often more than an inspector. He can be a goodwill ambassador for the PM program.

What he says and does and how he goes about it can be critical to the PM program. Typically, he should drop in at the start of his inspection and let the operating personnel know why he is there, and what he will be doing.

Often this preinspection visit is worth the time. Operating supervisors may have some real concerns about their equipment and are anxious to point them out. An understanding PM inspector—by merely listening to these problems—can help make the PM program a success. He can gain cooperation and credibility.

During the inspection, the PM inspector should talk to operators and learn of their concerns, to be made aware of problems.

As the PM inspection is completed, the inspector should make a special effort to reassure operating personnel that the problems they pointed out were looked into, solved, or that they require further attention.

The PM inspector is the maintenance department's representative. Often overconscientious inspectors attempt to fix everything as they go. In violating the basic detection-orientation idea, they undermine the PM effort. Maintenance supervisors must be aware of this tendency and control it.

The Maintenance Foreman

The maintenance foreman is the key to the whole PM effort. He gets it started. The inspectors are usually his men. Deficiencies come back to him for decisions. He advises the operating supervisors on repair decisions. He briefs the maintenance planner on requirements for major jobs so that details can be taken care of. Often he sees that follow-up repairs are made. Essentially, the PM program's success hinges on the maintenance foreman.

In smaller maintenance organizations, the foreman performs a dual role. He must evaluate reported deficiencies, then plan for and schedule follow-up repairs. Under these circumstances his role is even more important in PM.

The Maintenance Planner

In larger maintenance organizations with planners, every major job developed by the PM inspection program should flow through the maintenance planner. To ensure that repairs are made promptly, the planner must work closely with the foreman. Once the task is understood, materials must be gathered, labor estimates made, fabrication completed, and coordination with equipment operators carried out. When these actions are completed, the job is ready to be scheduled.

The planner and the foreman must work together as closely as two members of a team in a relay race. A successfully run first lap by the planner—job planning—allows the foreman to execute a well-planned job effectively in the second lap.

The PM Inspection Effort in Perspective

Preventive maintenance inspections are pivotal to the whole PM program. They uncover problems in time to ensure repairs before failure. Thus, not only are emergencies avoided, but downtime is reduced and more deliberately planned work can be done.

The increase in planned work means that better use can be made of maintenance labor—it works more effectively.

There is no magic to the development and implementation of a PM inspection program. Its success depends on the joint effort of operators and maintenance and the hard work of foremen, inspectors, planners, and maintenance managers.

THE DECISION TO USE A COMPUTER FOR PREVENTIVE MAINTENANCE SCHEDULING

The main advantage in using the computer to schedule PM activities is its ability to handle a wide variety and large number of PM services. There can be a multitude of services, many at different frequencies. The administrative task of sorting out each of these services to ensure each is properly scheduled can be considerable.

The simplest and most effective use of the computer is to produce a preventive maintenance work order. In turn, the PM work order is issued to maintenance personnel who will carry out the PM service.

By putting the PM program on the computer, a number of advantages accrue:

- A record can be made of each PM service performed to create a historical documentation.
- Labor recorded against PM services can be accumulated to show the amount of manpower required to sustain the PM program.
- Deficiencies attributed to PM inspections can be summarized to reflect the effectiveness of the PM program.

Availability of computer services provided by one's own organization can be a limiting factor if computer time or programming services are not available on a timely basis.

Use of the computer for scheduling PM services may have to be fitted in with other department priorities.

If such priorities for the use of the computer are placed ahead of scheduling PM, it is useful to explore other sources of computer support. Alternate sources include time sharing and use of the mini-computer. Each of these alternate sources has the limitation that it is not integrated with the company's total data processing network. When such is the case, it is better to acknowledge these limitations and keep the momentum of the PM program going. For example, this may mean short, coded PM messages, rather than detailed checklists.

Yet, PM programs usually require a trial or testing period. Therefore, interim use of time sharing or use of the mini-computer can provide opportunity to gain experience while waiting for more desirable, in-house computer support to become available.

In making a decision to use the computer to schedule PM actions it is useful to review the entire information system. By doing this, the type of information needed from the information system as it relates to PM can be seen more clearly. See Figure 2-12.

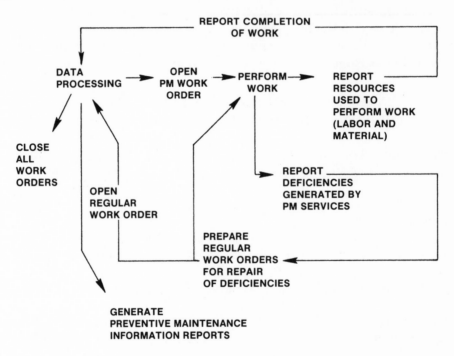

Figure 2-12.

The PM work order is used primarily to specify a particular PM service. However, additional work orders are prepared when PM services, particularly equipment inspections, cleaning, and testing, are conducted. These work orders are then fed back into the overall work order system.

Generally, specific information on preventive maintenance is included with other information. Typical information includes:

- *Labor Utilization*—Manpower used to perform maintenance is usually subdivided into the type of work performed. These types might include: preventive maintenance, scheduled maintenance, unscheduled repairs, and emergency repairs. A percent distribution of manpower use is shown to indicate how much manpower was used on PM and whether or not such use has helped to reduce unscheduled and emergency repairs.

- *Repair History*—Specific repairs recorded under repair history are attributed to PM services such as inspection, testing or cleaning to reflect the effectiveness of PM.

- *Cost*—The relative cost of PM-related activities is shown as part of a cost report. This data relates PM costs to overall maintenance costs. In turn, judgments can be made regarding the effectiveness of PM in reducing maintenance costs.

When weighing the decision to use the computer for scheduling PM services, the following guidelines are important:

- Try to use in-house computer services in order to have a totally integrated work order system.

- If in-house computer services are not available, examine other options such as time sharing and use them so as not to lose the momentum of the PM program.

- Always review the PM services planned for use with the computer against the overall work order system. This step ensures that essential information on PM will be made available.

PREDICTIVE MAINTENANCE (NONDESTRUCTIVE TESTING)

Predictive maintenance fits within the broad definition of preventive maintenance as a further means of prolonging equipment life and preventing premature failures.

Predictive maintenance is often referred to as nondestructive testing. Generally, it provides a group of techniques, whereby impending failures can be predicted. Various techniques pinpoint specific components that are performing abnormally and are likely to fail. Likewise, these techniques are able to isolate those components that are performing normally and do not have to be replaced.

The diagnostic tools that are available include:

- *Vibration Analysis*—vibration is measured at various critical points of rotating equipment. Variations noted from a previously recorded norm or baseline help to isolate the specific problem or component in need of replacement.

COMPARING TECHNIQUES OF
PREVENTIVE AND PREDICTIVE MAINTENANCE

Maintenance Action	Preventive Maintenance	Predictive Maintenance
Diagnosis and Inspection	Repair history is used to predict estimated equipment life. Physical inspection and standard PM services are used to prolong life of equipment.	Diagnostic tools are used to compare current or continuous operation against normal operating profile.
Detection of Abnormal Operation	Visual inspections. Teardown inspections. Evaluation of repair history. Estimation of theoretical failure time.	Monitoring during operation. Automatic warning of failure. Early detection.
Correction of Deficiencies	Periodic overhaul with possible replacement of sound components.	Spot replacement of only unsound components.
Cost of Repair	Overstocking of parts for worst situation. Budgets based on historical repairs.	More accurate inventory levels. Costs evaluated in detail based on predicted data.

Figure 2-13.

Generally, preventive maintenance techniques are broader and less sophisticated than the predictive measures. Certain advantages in preciseness occur through correct application of predictive techniques.

- *Dye Testing*—the use of dyes to help spot cracks not otherwise visible.

- *Sonic Testing*—the use of sonic equipment to measure thickness on certain critical wear points. For example, a pumping station handling abrasive slurry tends to rupture when fast-moving slurry wears piping at sharp turning points. Sonic testing would reveal excessive wear and alert to the need for replacing the piping at that point.

- *Acoustical Analysis*—sound measurements taken repetitively at fixed locations in order to detect abnormal noise level variations.

- *Infrared Analysis*—location of unusual "hot spots" using infrared techniques.

- *Other Techniques*—moisture absorption, flow or temperature measurements, or chemical analysis.

The techniques of preventive and predictive maintenance equip the maintenance department with a wide range of diagnostic tools. Figure 2-13 compares preventive and predictive measures to aid in selection of the most appropriate technique.

COMPONENT REPLACEMENT AND PERIODIC MAINTENANCE

Experience coupled with repair history will confirm that many components have a specific wear-life. Once this is known, the time span between previous instances of component replacement and current elapsed time provides opportunity to project and plan the next replacement.

This is known as periodic maintenance. It prescribes that certain actions be taken when specific time periods have been passed. Normally, this type of maintenance is best handled by the maintenance planner. A forecasting technique can be used to mark the intervals between these specific events. As the event approaches, the planner takes steps to plan for the event. This forecasting is best handled by the planner because he can bring the necessary continuity to it. See Figure 2-14.

The primary source of periodic maintenance data is equipment repair history. Simply stated, an equipment repair history is a chronological listing of all repairs that have been made on critical production equipment. The Chinese proverb that says, "The palest ink is better than the best memory," is particularly appropriate for maintenance managers engaged in the constant battle against chronic equipment breakdowns. Maintenance person's memories are a poor substitute for written equipment repair histories.

However, several questions need to be answered regarding repair history. For example:

- Why should an equipment repair history be kept?
- Which equipment is considered critical?

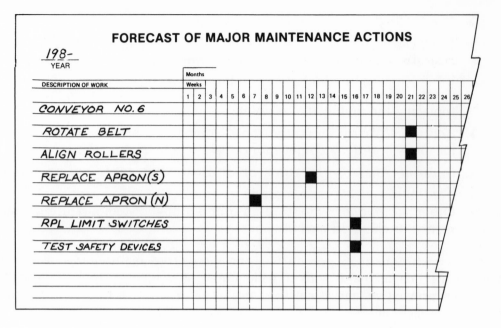

Figure 2-14.

The Forecast of Major Maintenance Actions provides the planner with a focal point for all periodic maintenance activities. While these are often component replacements, other activities can also be included.

- How should equipment repair history be collected?
- What should be done with the repair history?

An equipment repair history program is a form of "intelligence" that provides the maintenance department with the means to identify, classify, and correct persistent equipment problems. An effective equipment repair history system and an effective maintenance operation usually go hand in hand.

A comprehensive equipment repair history is as necessary to the success of maintenance operations as a preventive maintenance program is. The two differ in that a PM program gives the plant engineer guidelines for determining what equipment maintenance is necessary and an equipment repair history tells what has happened to the equipment, why it happened, and what it means.

Why Should Equipment Repair History Data Be Collected?

The pattern of significant equipment repairs such as machinery rebuilds, overhauls, and major teardown inspections over a period of operating time can indicate if changes in the maintenance routine are necessary. For example, the pattern of repairs will indicate:

- if equipment is undermaintained or overmaintained.
- if equipment should be modified.
- if overhauling or rebuilding is needed.
- if the components should be changed at different intervals.
- if sufficient lubrication is being applied.
- if operators are abusing the equipment.

The maintenance manager should make all decisions regarding changes in maintenance actions for each piece of equipment.

Generally, the maintenance manager is interested in the similarity between current repair needs and the actions shown in repair history. Most foremen view the history only as a diary of repairs rather than as a means of determining a pattern of equipment failure. Often, the foreman is too pressed by current problems to use the repair history to analyze the reasons behind equipment repair problems. The maintenance manager will be more inclined to use repair records as an analytical tool to spot persistent repair problems.

What Is Critical Equipment?

Two criteria determine whether equipment is critical: (1) cost to maintain it, and (2) importance in the process or use of the equipment. Maintenance costs of equipment can vary widely. Not all machines produce a consistent cost trend.

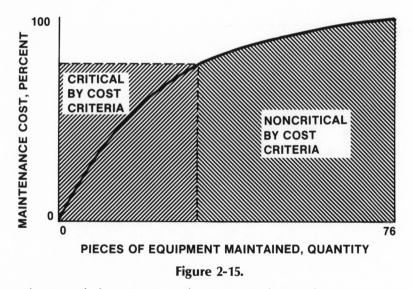

Figure 2-15.

In this typical plant situation, about 33 out of 76 machines accounted for 75 percent of the equipment maintenance costs. Graphs of this type help identify critical equipment from a maintenance cost standpoint.

Because of the variance in costs, judgment of equipment criticality must not be made with only one year's cost data; several years should be charted.

The results of a critical equipment cost analysis can point out which units of equipment should be subjected to repair history analysis. In most cases, about 20 percent of the total equipment population accounts for 75 to 85 percent of the total maintenance cost. See Figure 2-15.

A cost analysis will indicate which of the vital equipment is expensive or inexpensive to maintain.

The next step that must be taken is to evaluate the importance of the equipment:

- Is it one of a kind?
- Is it the connecting link between several important steps?
- Is it a necessary sensory or measuring device?
- Does it require sensitive adjustments or continuous monitoring?
- Can it be operated only by highly trained personnel?
- Can it be bypassed?

An assessment of the actual criticality of the equipment can be made only after this examination has been completed.

How Should Repair History Be Kept?

Equipment repair history should be kept where maintenance is being performed. The maintenance foreman should be responsible for keeping the information up-to-date if a manual log is used to record repair history. Regardless of the data-gathering method used, the information included in the log must include a chronology of significant repairs for analysis. Figure 2-16 shows a repair log form that has been used successfully.

Note that repairs are listed chronologically, but there is also room on the form to record information important to the foreman.

What Should Be Done with Repair Data?

Equipment repair history is valuable in identifying persistent maintenance problems. Repair history will indicate that a piece of equipment should be reengineered to correct a problem. Components that consistently fail after a specific wear-life period can be replaced before they fail and cause costly shutdowns. As plant equipment gets older, the interval between overhauls will usually change drastically. A repair history will indicate when machines should be overhauled to avoid an unexpected failure.

Repair histories should include a failure coding system in which a number corresponding to the reason for failure is entered in the log. The failure pattern

Repair History Log	BELT CONVEYOR NO.16 516482 / 175 FT INCLINE		
DATE	REPAIR ACTION	FAIL CODE	REMARKS
2/17/80	ADJUST GUIDE ROLLERS	O	
3/27/80	LUBE SUPPORT ROLLERS	3	
4/24/80	REMOVE DEBRIS UNDER SUPPORT ROLLERS	1	
5/11/80	REPLACE TORN BELT SECTION	4	8 FT
6/16/80	REPLACE BELT LACING ON 5/11/80 JOB	8	
6/16/80	REMOVE DEBRIS UNDER BELT	1	WARNED OPERATOR
7/2/80	REPAIR LOCK-OUT SWITCH	1	
8/11/80	REPLACE 3 GUIDE ROLLERS	9	PART NO. 8X-7114-2081
9/12/80	CLEAN UP PRODUCT SPILLAGE	7	WARNED OPERATOR
10/13/80	REPLACE DAMAGED SUPPORT ROLLER	6	
10/18/80	REPLACE TORN BELT SECTION	7	16 FT

Figure 2-16.

This repair history log is typical of the approach used by maintenance managers to record repair actions in a chronological manner. Note that there is also room on the form for the foreman to record other items he feels are important.

shown by the coding system helps to identify the nature of corrective actions. For example, see Figure 2-17.

Failure Code			Percent of Instances
1	☐☐☐☐	**Operator error** ☐☐☐☐	16
2	☐☐☐☐	**Equipment abuse** ☐☐☐☐	11
3	☐☐☐☐	**Lubrication failure** ☐☐☐☐	2
4	☐☐☐☐	**Blockage** ☐☐☐☐	20
5	☐☐☐☐	**Fire** ☐☐☐☐	3
6	☐☐☐☐	**Accident** ☐☐☐☐	28
7	☐☐☐☐	**Product spillage** ☐☐☐☐	11
8	☐☐☐☐	**Improperly repaired** ☐☐☐☐	6
9	☐☐☐☐	**Material fatigue** ☐☐☐☐	1
0	☐☐☐☐	**Other** ☐☐☐☐	2

Figure 2-17.

In this illustration, the high percentages of failures caused by operator error, accident, and blockage show a need for extensive operator training and increased supervision.

SUMMING UP

The preventive maintenance program can prolong equipment life and prevent premature equipment failure. Yet, there are other less tangible benefits that accrue.

When done effectively, equipment inspection, testing, and cleaning can uncover deficiencies in the earlier stages of deterioration. Thus, the time available for making repairs before failure is greater. In many cases these deficiencies relate to major jobs that require planning before they can be performed. Therefore, the greater lead time before work must be done can often be used to plan major work. This, in turn, creates opportunity to perform work on a planned basis. Herein lies one of the principal advantages of PM.

As more work is planned, better use is made of maintenance labor resources. When using maintenance resources for similar jobs done under planned versus unplanned conditions, material costs are about the same. Generally, savings of 12 to

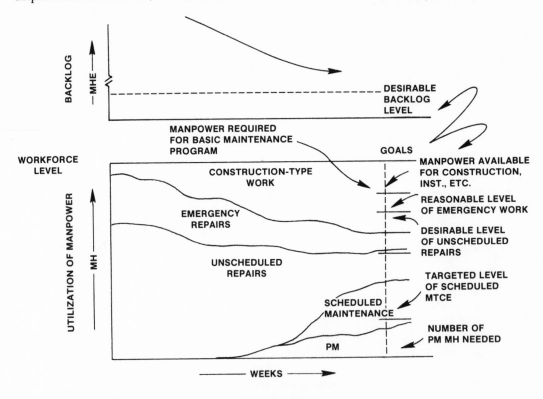

Figure 2-18.

As the number of man-hours used for PM services increases, the amount of unscheduled and emergency work decreases. More deficiencies are found sooner and more work is able to be planned. There is an increase in scheduled maintenance and better manpower use. Often a surplus of labor is realized and more construction-type work (nonmaintenance) is able to be done. Simultaneously, the backlog is reduced.

15 percent in labor are possible in performing similar planned jobs versus unplanned jobs.

In turn, this aspect of labor savings applies to the whole maintenance program. See Figure 2-18.

For those who have installed successful PM programs, the rewards are well known. This chapter is intended for those who have yet to do it.

Perhaps the best advice for those who are committing an effort to PM is the need for planning the program before getting underway. The most important step is to secure the necessary commitment from management, equipment users and maintenance. Developing and installing the program, and then making it work, is never easy. However, it is easier with this commitment. With this commitment comes the help—outside of maintenance—that can make the program succeed.

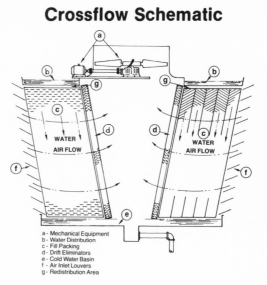

Crossflow Schematic

a - Mechanical Equipment
b - Water Distribution
c - Fill Packing
d - Drift Eliminators
e - Cold Water Basin
f - Air Inlet Louvers
g - Redistribution Area

Cooling Tower Technology

Robert Burger

SCOPE OF CHAPTER

To maintain a cooling tower, or any mechanical equipment, the operator must be aware of the theory that has been developed to obtain the desired results. He must clearly understand the various elements of the machinery, plus their relationships and functions to each other, so that the device can be utilized in a professional manner at maximum levels of efficiency.

Therefore, this chapter on cooling towers consists of three sections:

1. Maintenance

2. Theory

3. Application

Since the purpose of the textbook is *maintenance*, this portion appears first, followed by theory and application for those who desire to obtain better understanding and utilization of their equipment.

Cooling towers are manufactured in a great variety of sizes and shapes ranging from small five-ton factory-assembled packages to field-erected units measuring up to, and sometimes exceeding, 150 feet long by 50 feet wide by 50 feet high. The service of cooling towers can vary from air conditioning applications, where the water is put on the tower at 95° F (degrees Fahrenheit) for mechanical refrigeration, to 102 to 105.5° F for absorption equipment, returning in most cases from the tower at 85° F. The wet bulb, depending upon the area of the country, is between 65 to 80° F. Chemical process and power generation temperatures run the gamut of 95 to 150° F on the tower discharging at 83 to 90° F off the tower, where the wet bulb again varies depending upon the geographic location.

Acceptable limits of operation are determined by the requirements of the installation. It is obvious that when the mechanical equipment breaks down, it is past its acceptable limits. When the water is returned 2 to 5 degrees or more warmer than manufacturer requirements for the compressors and condensers, corrective actions must be taken, to avoid excessive power use and to keep the equipment on the line. Again, strict guidelines cannot be published since each individual case is determined at the site by the operating requirements.

HISTORICAL DEVELOPMENTS

The modern cooling tower developed in an orderly progression. In the "dark ages" of plentiful resources and the absence of ecological conservationists, cooling water was used merely "once through," discharged, and forgotten. Where topographical considerations were analyzed and available, large ponds or lakes (artificial or natural), or canals were utilized to hold, cool, and recirculate the water.

To facilitate cooling and reduce the amount of real estate used, spray systems were installed to aerate the water in the ponds and canals. This obtained faster cooling; not by relying on top layer evaporation and sensible heat exchange, but by offering more water surface to the atmosphere in the form of spray for evaporative cooling.

In order to obtain better control, it was soon determined that by rotating the nozzles 180 degrees and spraying down in the box instead of up, lower temperatures could be obtained. Soon, instead of being dependent upon prevailing winds, fans or air movers were designed to master and assist nature's unreliable aerodynamics.

As more was understood of the mechanics and hydrodynamic process of water cooling, fill packing was added to slow the vertical fall of water and to provide a longer air/water interface contact for greater levels of cooling and heat transfer.

Today, all of these techniques are used, however, some are used in modified forms.

THE COMMON DENOMINATOR

The New Orleans Superdome air conditioning/refrigeration system is rated at 10,000 tons, circulating 30,000 gallons per minute (GPM) of water, while a small-town shoe store cools its premises with a 10-ton central air conditioning system circulating 30 GPM. There is one common denominator.

The fabric of our society would collapse without electric generating plants, whether they be one megawatt (MW) installations, for a small apartment house complex burning fossil fuel, or 1,000 MW nuclear steam-electric facilities for a large public utility. Again, there is one common denominator.

Chemical plants throughout the country yield an unimaginable number and variety of products that our civilization utilizes. In all of our activities, day and night, we rely upon them, and in these chemical operations, there is one common denominator.

This common denominator cannot be ignored: It is <u>waste heat</u>.

In the above examples, waste heat is absorbed by water having a temperature colder than the process. This hot water must now either be discharged into a body of water or cooled and recycled. The common alternative for all systems is to lower the water temperature in a device called a cooling tower. Here, the waste heat is rejected into the atmosphere, and the colder water is then recirculated throughout the system, thereby conserving our water resources.

The problem is enormous, since industry uses approximately 500 billion gallons of water a day: which, if not recirculated, would seriously deplete the available water supply. The unwanted heat from the water must be restored to the environment without creating ecological havoc or adding unnecessary expenses to the inflationary spiral.

The development of cooling towers solved the problem of possible ecological damage of returning hot water to the environment, and enabled industry to recycle the greater proportion of the water it required at reasonable capital investment costs.

Operating procedures, costs, and upgrading potential for greater efficiencies must be investigated constantly to provide economic returns for the installation. To optimize the upgrading of the tower, we must first understand the internal elements and their functions. A comprehensive maintenance program must also be instituted to obtain maximum equipment utilization.

COOLING TOWER MAINTENANCE AND INSPECTION

Properly functioning cooling towers make money by producing colder water for refrigeration, cooling, or process water. Colder water returning from the cooling tower, under similar conditions of load and temperature, can add up to many dollars saved in energy consumption and/or product results. In order to keep a tower functioning efficiently, all of its components must be put in optimum working order so that excessive energy charges and performance slippages are eliminated.

Cooling towers are mass heat transfer devices that also consume energy in the course of their performance.

Refrigeration equipment can utilize over 450 horsepower for motors to drive compressors; or steam is utilized to power the turbines that consume energy and money to operate. A typical 1,000-ton refrigeration system (circulating 3,000 GPM of water) can use over $350,000 a year in electrical and/or steam energy.

Chemical plants manufacturing a saleable condensate can turn out over $1,000 a day in additional materials, which will more than pay for an extensive upgrading, rebuilding, or maintenance program of large industrial cooling towers.

Power generating plants use their electricity to generate power, and the degree of consumption is called the energy penalty. Again, colder water uses less energy and reduces the penalty, thereby producing more saleable electricity for the customer.

Obviously, it is necessary for the operator to carefully inspect his tower to eliminate any deficiencies that prevent optimum cold water temperatures returning off the tower to the equipment, to make note of these problems, and to budget money plus time to effect repairs.

There are many ways to obtain better performance from a cooling tower:

1. Inspection

2. Maintenance

Over the years the equipment loses some of its efficiency because of wear and tear, misalignment of parts, and general deterioration that occurs when inspection and maintenance are not effectively provided.

The oldest and best tools available to every maintenance department are its personnel's *eyes*, Figure 3-1. Proper inspection and adequate attention paid to discrepancies will more than pay for themselves in a smoother, more efficient operation.

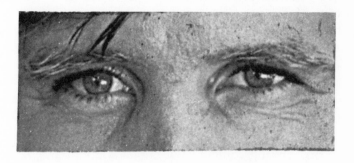

Figure 3-1.

The best inspection tool ever found and available to all maintenance personnel is a person's eyes.

When the cooling season draws to a close, operators of air conditioning and refrigeration systems plan shutdowns and schedule normal maintenance of the compressors, turbines, shells and tubes. The cooling tower, an important element of the entire system (sometimes because of its location) is often ignored until "five minutes" before turn on. Industrial towers, due to the nature of the operation, are never turned off except for scheduled turnarounds that are usually spaced many months apart.

Many owners and large industrial users rely on periodic inspections and reports by a professional field engineer, who spends at least one to two days at the site and submits a detailed evaluation of discrepancies noticed and suggested repair programs, together with the possible capability of thermal upgrading.

This section discusses the procedures that parallel the methodical manner in which field inspections are made. While there are a great variety of cooling tower types, the elements of all units are the same, and an individual checklist can be worked out covering all items as a preplanning procedure for developing a comprehensive inspection program. Figure 3-2 is typical of large industrial installations, but can easily be modified for any specific installation, whether it be a 50,000 gallon per minute crossflow or 50-ton (150 GPM) blow-through counterflow.

There are three major divisions of a cooling tower: exterior structure, mechanical equipment, and interior structure. We will investigate each division as a unit, marking down discrepancies as they appear.

1. *EXTERIOR STRUCTURE:* This inspection is the visual walk-around examining the following components.

   A. *Louvers:* Figure 3-3. Determine the number of missing louvers. Measure accurately the existing louver size, length, and width, and measure the thickness very carefully, since a 1/16" difference can prevent the new louver blades from fitting in the existing support slots. Out of position louvers should be checked for repositioning. If the louvers are wood, look for splits or breakage and check for wood rot and decay. Examine asbestos, cement, or fiberglass reinforced polyester louvers for breaks, cracks, and corner damage, which can prevent proper alignment. Examine wood louver supports for rot and decay; if made of galvanized steel, observe the condition of the protective coating and make notes of broken and rotted supports. See that attaching hardware is adequate and properly supports the louver posts. Examine louver supports for ice damage.

   B. *Casings:* Figure 3-4. On corrugated or flat cement asbestos board, transite, or fiberglass reinforced polyester casing, look for cracks, leaks, condition of attaching hardware, and other physical damage. Corner rolls should be observed for integrity and security of attachment. On wood casing towers, look for loose boards, evidence of wood decay, leaks, and wood rot. Wood for casing of towers is a rather old technique and usually consists of a double layer. This should be checked from the interior to determine if wood rot dictates entire replacement with modern fiberglass sheathing.

   C. *Access Doors:* Check for missing doors, broken or loose hinges or latches, general conditions of doors in operation, and any evidence of physical damage. Measure doors accurately if replacement is necessary.

COOLING TOWER INSPECTION CHECKLIST

Robert Burger Associates, Inc.

RBA Job No. _____

Owner _____ Date Inspected _____

Plant _____ Inspected By _____

Location _____ Tower Manufacturer _____

Owner Designation _____ Installed _____ 19 _____

Water Treatment Used _____ Model No. _____

Design Conditions _____ GPM _____ HW _____ CW _____ WB

Condition: 1-Good; 2-Repair; 3-Replace	1	2	3	Condition: 1-Good; 2-Repair; 3-Replace	1	2	3
EXTERIOR STRUCTURE:				MECHANICAL EQUIPMENT			
1. Endwall Casing & Access Doors				21. Drive Shafts (Type) _____			
2. Louvers (Type)_____				22. Speed Reducer			
3. Drain Boards _____				Series _____ Ratio _____			
4. Stairway _____				Oil Level _____			
5. Fan Deck _____				Oil Seals _____			
6. Fan Deck Supports _____				Vent _____			
7. Handrails _____				Back Lash _____			
8. Ladders & Walkways _____				Pinion Shaft Play _____			
9. Distribution System _____				Fan Shaft End Play _____			
Headers (Type) _____				Last Oil Change (Date)_____			
Distribution Basin _____				Oil Used _____			
Water Level _____				23. Fans			
Flow Control Valve (Size)_____				Dia. _____ Type _____			
Nozzles (Size) _____				Hub _____			
Water Distribution _____				Blades _____			
10. Spray System & Spray Nozzles _____				Hub Cover _____			
11. Fan Cylinders (Type) _____				Tip Clearance _____			
INTERIOR STRUCTURE:				No Vibration _____ Vibration _____			
12. Fill (Type) _____				Additional Components (If installed on tower)			
13. Columns _____				Fan Guards _____			
14. Girts				Oil Gauge & Drain Lines _____			
15. Diagonals _____				Vibration Limit Switches _____			
16. Partitions & Doors _____				Other: _____			
17. Eliminators (Type)_____				_____			
18. Walkway				_____			
19. Cold Water Basin (Type)_____				_____			
Water Depth _____				24. Motor: Mfr. _____			
20. Mech. Equip. Support (Type) _____				Name Plate _____ HP _____ RPM _____			
				Phase _____ Cycle _____ Volts _____			
				Amperes _____ Frame _____			

REPLACEMENT PARTS REQUIRED:

QUANTITY	DESCRIPTION	ORDER FROM	DATE REQ'D

MAINTENANCE WORK REQUIRED: DESCRIPTION | REQ'D COMPLETION

_____
_____
_____
_____

(Use back of this sheet for additional requirements or notes.)

Figure 3-2.

A checklist should be developed for each individual installation to ensure proper coverage of all items.

Figure 3-3.

Air intake louvers should be checked for missing or broken blades.

Figure 3-4.

Casing walls should be inspected for leaks, cracks, or deterioration.

D. *Distribution System:* Observe flange connectors at top of riser for security and gasketing and make notes of effectiveness of corrosion control coating on supply lines and risers. On counterflow towers, check where header enters casing to determine if it is properly caulked and no apparent leaks exist. On crossflow towers, (Figure 3-5) look for deterioration in distribution basins, splash guards and associated piping. With water trough configuration, check boards for warpage or splitting on basin sides and leakage caused by gaps in trough.

Figure 3-5.

Hot water distribution basins can be checked for debris in metering orifices and missing nozzles, together with integrity of the structure.

E. *Drain Boards:* Look for evidence of water being diverted outside the drain to tower exterior. Check for displaced or damaged drain boards. If drain boards are metal, observe for corrosion and rot. Look for missing boards and condition of fasteners on all types of materials.

F. *Stairways:* Note deterioration or loose handrails and kneerails. Check stringers and stair treads for missing components, breakage, and cracks.

Examine structure for looseness, evidence of rot, and condition of fastening devices. Steel stairways should be checked for evidence of corrosion or acid attack. All fasteners should be examined for security (in most installations, it is prudent to shake the stairway before using). Special attention should be paid to condition and security of upper braces of cantilevered platform supports.

G. *Ladders and Walkways:* Before climbing any ladder, check for security of fasteners by vigorously shaking ladder rails, then make sure that all the rungs are in position. Note missing or deteriorated rungs. Check treads on walkways for security, rot, or corrosion. Check stringers and rails for corrosion. In order to comply with O.S.H.A. safety requirements, any ladder over 20 feet must have a cage or safety device to protect the climber. Further, a cage above 20 feet must have welded platforms at 20 feet intervals or be protected by a safety device. Check these appliances for security and evidences of corrosion, rot or weakness. An old surviving "ladder climber trick" is to grasp ladder rails rather than rungs, since rungs could be rotted and weak.

H. *Fan Decks:* Examine fan decks for decay, missing or broken members, (Figure 3-6), and gaps between boards that can cause short-circuiting of air circulation. On steel decks check for corrosion, holes, and condition of corrosion control coating. Fan deck supports should be examined for decay, loose or broken members, and condition of fastening devices.

I. *Fan Cylinders:* See Figure 3-7. Make sure that fan cylinders are securely anchored to the fan deck supports. Observe for looseness of fan cylinder and condition of fastening devices. Look for damage, missing structure parts, and deterioration of components. Observe wood rot, or where applicable, corrosion of steel sides, rings, or support strap banding. Inspect for proper tip clearance between fan blade and interior of the cylinder walls. O.S.H.A. safety requirements dictate a protective fan guard on cylinders less than five feet high. If a guard is installed, check condition of components for corrosion, sagging, and missing parts.

2. *INTERIOR STRUCTURE*

A. *Distribution System:* Inspect metal piping headers for decay, rust, or acid attack. Check flange connections for tightness and condition of fasteners and gaskets. See that all branch arms are secure; note any corrosion on flange connectors. Check for missing nozzles and, if practical, observe spray pattern (Figure 3-8) to see if nozzles are in good operating order and are not clogged. For crossflow water distribution, make sure that nozzles are cleaned and operating, and, if nozzles are not used, check to ascertain that all distribution holes are clear and are not clogged. Note condition of redistribution section beneath hot water basin for necessary repairs or replacements. If piping is fabricated

Figure 3-6.

Fan decks should be checked for security so that air does not bypass through leaks. Weakened fan decks are also dangerous for personnel walking on them.

Figure 3-7.

Destructive corrosion can move metal into the path of rotating fan blades.

Figure 3-8.

Spray pattern on fill and in cold water basins is an indicator of nozzle performance.

from wood, check condition of retaining bands, look for warpage or split boards, and check for excessive leakage. Water trough distribution should be observed for condition of wood and fasteners, for excessive gaps between sides and bottom of trough, and for alignment of splash cups.

B. *Mechanical Equipment Supports:* Inspect the mechanical equipment supporting steel for evidence of corrosion, and condition of the fasteners, straps, or hangers. Look at wood structural members in contact with the steel supports to ensure soundness or evidence of weakness that could result in loss of structural strength. If springs or rubber vibration absorption pads have been installed, check conditions of springs, mounting and adjusting bolts, corrosion of ferrous members, and condition of rubber pads. Check for iron rot where steel and wood are in contact. See Figure 3-9.

C. *Fire Detection:* Test the valves and the operating condition of wet or dry sprinkler system. Check for evidence of corrosion on pipes and connectors. See to it that activating devices are in apparent good order. If a thermocouple wiring system is used as an alarm indicator, check for broken wiring.

D. *Eliminators:* Check drift eliminators for broken, displaced or deteriorated blades, and for gaps or misalignment that will permit excessive drift. (See Figure 3-10.) Check supports for breakage or

Figure 3-9.

Where wood and iron come into contact, iron rot is a possible cause of deterioration.

Figure 3-10.

Drift eliminators deteriorate and droplets of water are drawn out of the tower by the fan.

deterioration. Look for clogging caused by algae, debris, or slime. Split or expanded crossflow fiberboard or plywood vertical support members should be examined for possible replacement.

E. *Fill:* Inspect tower fill for any breakage, deterioration and misplaced or missing splash bars. (See Figure 3-11.) Look for damage to splash bars supports and fill supports. If sections of fill have collapsed, look for damage from icing. Obtain design conditions of water and operating temperatures so that thermal calculations can be made later to determine degree of upgrading capability available by changing fill to a more efficient design. Fireproof asbestos fill should be considered as an alternate to expensive automatic deluge system, since piping and controls of a fire alarm system need continual checking and maintenance while, on the other hand, asbestos fill, by the nature of its fireproof construction, does its fire protection job without further attention.

Figure 3-11.

Worn fill splash bars create gaps and very inefficient cooling.

F. *Interior Structural Supports:* Columns, girts, and diagonal wood members can be tested for soundness by striking with a hammer. A high-pitched, sharp sound indicates good wood, while a low-keyed, dull sound indicates softness of wood. Wood rot will soften wood, so, if an area indicates possible rot, probe it with a screwdriver to further determine condition of the wood. Check carefully around metal fasteners for iron rot on wood in contact with metal. Check steel internals for corrosion and deterioration caused by rusting. Also, check bolt condition and tightness on both wood and steel interiors.

G. *Partitions:* Examine bolts and nuts in partitions for looseness and corrosion. Inspect for deteriorated or loose partition boards. Note if partitions go from basin to roof deck dividing individual cells so as to prevent windmilling of idle fans or short-circuiting of air between cells. See that wind walls parallel to air intake louvers are in position, and that boards or transite members are securely fastened. Check supports, either wood or steel, for condition of rot or corrosion and condition of the protective coating.

H. *Doors:* Make notation of missing doors, condition of hinges and fasteners, and look for corrosion, rot, or deterioration of door material. If cell partition does not have doors at basin level, indicate so on report.

I. *Cold Water Basins:* For wood basins, check deterioration, warped or split basin sides, open joints, and soundness of wood. (See Figure 3-12.)

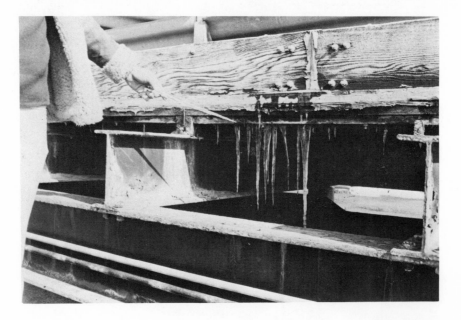

Figure 3-12.

During winter operations, leaks in basins are quite discernible.

Inspect steel basin for corrosion, holes, and condition of protective coating. Concrete basins should be inspected for cracks, breaking joints, and acid attack. All sumps should be observed for accumulation of debris, condition of screens, anti-turbulator plates, and freedom of operation of drain valves. If the basin is one continuous unit, partitions between cells, together with necessary sumps and piping, should be recommended. This adds flexibility for utilization of equipment by operating only the number of cells required for the heat rejection load, thereby conserving power.

3. *MECHANICAL EQUIPMENT*

A. *Fans:* Check fan hubs for corrosion. Inspect hub covers for corrosion and condition of attaching hardware. Inspect blade clamping arrangement for tightness and corrosion. Look at fan blades carefully for corrosion and erosion. Check fan blade pitch with bubble protractor for uniformity; check blades for build-up of solids that would change blade moment weight or air foil characteristic, thereby causing excessive vibration.

B. *Gear Boxes:* Check gear reducers for proper oil level. (See Figure 3-13.) Inspect oil for moisture and sludge. Rotate input shaft by hand back and forth against gear tooth contact to feel for backlash. Check input pinion shaft bearing for wear, by attempting to move the shaft radially. Look at fan shaft bearings for excessive end play by applying force up and down on a fan blade tip and note the movement of output shaft. A running clearance is built into some output shafts, which should not be confused with excessive end play.

Figure 3-13.

Oil levels can be determined by external dip stick or looking at the sight glass on the gearbox.

C. *Power Transmission:* Inspect drive shaft keys and set screws; check assembly hardware for tightness. Examine drive shaft couplings for corrosion, wear, or missing elements. Look at exterior of drive shaft visually for corrosion, and check interior by lightly tapping draft shaft for sound along its length for dead spots that could indicate internal corrosion. Carefully observe flexible connectors of both ends of shaft. A shaft guard is imperative to prevent fan wrecks. The weakest link in the power transmission train is the coupling connectors. Invariably when a coupling breaks, the rotating shaft swings upward and wrecks the fan blades, plus possibly doing damage to the gear box and motor. This can be prevented by the installation of simple, inexpensive shaft guards to absorb the upward thrust of the rotating broken shaft. (See Figure 3-14.) Belt-driven units should be observed for alignment, tension, and condition of wear of belts.

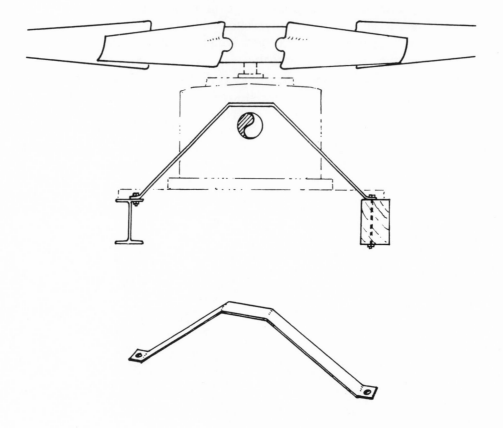

Figure 3-14.

Installation of a simple fan shaft guard by in-house maintenance personnel can prevent very expensive fan wreck.

D. *Motors:* Look for evidence of motor heating, and observe proper lubrication of the bearings. Excessive grease or oil applied to ball bearing motors can damage the unit by forcing grease into windings and causing deterioration of insulation. Look for deposits of dirt or dust at motor air intakes and check TEFC motors for conditions of air passages and fans. The drain moisture plug is supplied with motor; see if it is operational. Measure operating loads with amp probe and volt meter as compared with nameplate data; recommend pitching of fan blades up or down to compensate. Examine mounting bolts and attachments for security and corrosion.

All of the above items might seem excessively complicated and time-consuming. However, if a thorough investigation and description is to be made, it must cover all elements of the cooling tower. To facilitate the actual inspection, a checklist can be compiled (see Figure 3-2), which will greatly speed up the job and ensure that all aspects of the inspection were accomplished. This will also serve as a reference to writing and submitting your report.

Extending mechanical equipment life by proper maintenance, timely repairs, and intelligent utilization of colder water generated by a cooling tower rebuilt with modern engineering principles, are essential.

A routine preventive maintenance service checklist should be made for all of the equipment the Operating Engineer is responsible for. A three-card system could be put together for the cooling tower as follows:

1. *Prior to the start-up* of the cooling tower perform the following:

   A. Clean out tower basin.

   B. Caulk cracks and openings in the T&G joints with mastic.

   C. Tighten all bolts of columns, beams, connectors, and equipment.

   D. Clean, prime, and coat all areas of steel corrosion.

   E. Clean and lubricate electric motor bearings in accordance with the manufacturer's recommendations.

   F. Inspect and adjust gearbox oil level.

   G. Check shafts and couplings for security or tension and alignment of V-belts.

   H. Check starters, relays, coils, clean contacts.

   I. Operate fan driver and check bearings for overheating.

   J. Consult your chemical treatment firm for start-up.

   K. Fill system to basin level and check float and make-up equipment.

2. *During the cooling season* check tower under operation as follows:

   A. Observe fan and motor for unusual noise or vibration.

   B. Examine V-belts and/or shaft and couplings.

   C. Inspect basin water level, float, and bleed-rate.

   D. See that water pattern is uniform and that spray nozzles or metering orifices are not clogged.

   E. Check water treatment for proper operation.

   F. Clean intake strainer; check tower basin for accumulation of debris.

   G. Shut down equipment periodically to observe gearbox lubrication, fan bearing condition, or shaft coupling connectors for security.

   H. Observe steel casing and steel parts for rust and corrosion.

I. Perform any urgent repairs that cannot be scheduled when tower is shut down at the end of the season.

3. *At end of cooling season:*
   A. Clean and flush out tower leaving drains open.
   B. Drain only parts of tower subject to freezing.
   C. Clean drift eliminators, fill decks, louvers and spray heads of observable debris.
   D. Schedule repairs necessary before placing tower into operation.
   E. When tower is completely dried, clean, prime, and coat areas requiring it, using recommendations from a coating specialist.
   F. Fill gearbox with oil to prevent condensation.
   G. To protect the owner's investment in this tower, go back to the complete inspection checklist in Figure 3-2.

COLD WATER MAKES PROFITS—OPERATIONAL RESPONSIBILITY

Because of insufficient time and limited personnel, most maintenance managers pay a minimum of attention to the cooling tower, taking its function—to cool hot water—for granted.

Many managers are not even aware of, or do not consider, the role of the cooling tower in the company's profit or loss picture. It is the Operating Engineer's function to employ the equipment in an efficient a manner as possible to produce money and profits for management.

The major area of profit to analyze is the utilization of the cold water. In the refrigeration/air conditioning system, the colder the water, the less energy is required to power the equipment. In chemical product manufacturing plants, the colder the water, the more efficient the condensation and the greater the volume of saleable condensate product is yielded for the plant at lower cost. In power plants, colder water reduces the electrical generation penalty (the cost of electricity used in the process of producing saleable or usable power).

The second area of cost to consider is the electrical and/or steam dollar expenditure which is purchased to power the systems. The cooling tower is a high energy utilization machine as well as a mass heat transfer device.

After the cooling tower has been erected, proper operation and maintenance is required to yield the design conditions of cooling. By not maintaining towers adequately, they can lose appreciable performance as soon as they are turned on.

This chapter on cooling towers is aimed at one basic end result—COLD WATER.

DESIGN CONDITIONS

The criteria of cooling tower performance are outlined in the design conditions, (Figure 3-15), specified when the cooling tower is purchased and/or rebuilt to mean

DESIGN CONDITIONS

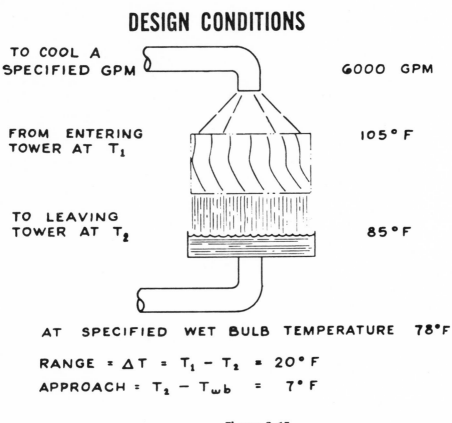

TO COOL A
SPECIFIED GPM   6000 GPM

FROM ENTERING
TOWER AT T_1 105°F

TO LEAVING
TOWER AT T_2 85°F

AT SPECIFIED WET BULB TEMPERATURE   78°F

RANGE = ΔT = $T_1 - T_2$ = 20°F

APPROACH = $T_2 - T_{wb}$ = 7°F

Figure 3-15.

Design Conditions in This Example Are to Cool:

a. 6,000 GPM of Water
b. Entering at 105°F
c. Leaving at 85°F
d. At ambient W.B. of 78°F

the cooling of a specific quantity of circulating water entering the tower at a particular temperature and leaving at a definite value. The difference between these two temperatures is called *The Range* or *Delta T*. This cooling of the specific amount of water is also to be performed at a wet bulb temperature that is noted in the specifications. The difference between the cold discharged water and wet bulb is called *the approach to the wet bulb*.

A 1°F, or 0.6°C, colder water returned to the compressors and condensers in air conditioning refrigeration equipment calculates, by use of Enthalpy Charts, a 3¼ percent savings in electrical energy input to these machines. Therefore, a little more than 3°F (1.7°C) colder water off the tower can save 10 percent of electrical energy and resulting charges thereof at any given time. A 2,000-ton (1,815 metric ton) refrigeration system circulating 6,000 GPM (22,700 liters/minute) could use $350,000 in electrical power a year. A 10 percent savings of $35,000 obtained by sending colder water to the machinery will be a significant factor towards rapid payback for the upgrading of the tower.

COOLING TOWER FUNDAMENTALS

The basic function of cooling tower operation is that of evaporative cooling and exchange of sensible heat. The air and water mixture releases latent heat vaporization. Water exposed to the atmosphere evaporates, and as the water changes to vapor, heat is consumed at approximately 1,000 BTUs per pound of water evaporated. The heat is taken from the water that remains by lowering its temperature.

However, there is a penalty involved—the loss of water that goes up to the cooling tower and is discharged into the atmosphere as hot moist water vapor (Figure 3-16). Under normal operating conditions this amounts to approximately 1.2 percent for each 10°F (5.5°C) of cooling range. Sensible heat that changes temperature is also responsible for part of the cooling tower's operation. When water is warmer than the air, there is a tendency for the air to cool the water. The air gets hotter as it gains the sensible heat of the water and the water is cooled as its sensible heat is transferred to the air.

Figure 3-16.
Cooling tower plumes (fog) are the byproduct of evaporative cooling resulting in loss of water volume.

Cooling due to the evaporative effect of the release of latent heat of vaporization amounts to approximately 75 percent, while 25 percent of the heat exchange in the cooling tower is sensible heat transfer. The cooling tower, like any other device, does not escape the unchangeable law of the indestructability of matter. A cooling tower is merely a machine that takes a mass of heat from one area and moves it to another area. In technical terms, it is referred to by thermal engineers as "the heat rejection solution" or "correction of the heat penalty generation of compression equipment." The cooling tower is a machine that moves heat from point "A" to point "B" and ultimately discharges this heat into the atmosphere.

COUNTERFLOW VERSUS CROSSFLOW

All cooling towers, except the small packaged specialty models, are custom-made. The manufacturer varies the shape, size, configuration, and input to meet the

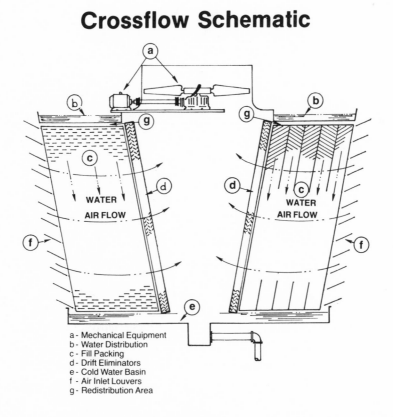

a - Mechanical Equipment
b - Water Distribution
c - Fill Packing
d - Drift Eliminators
e - Cold Water Basin
f - Air Inlet Louvers
g - Redistribution Area

Figure 3-17.

Crossflow schematic cooling tower indicating the various elements.

Counterflow Schematic

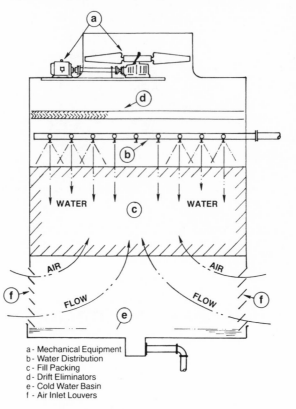

a - Mechanical Equipment
b - Water Distribution
c - Fill Packing
d - Drift Eliminators
e - Cold Water Basin
f - Air Inlet Louvers

Figure 3-18.

Counterflow schematic cooling tower indicating the schematic elements.

particular set of thermal parameters required by the customer. Since they are available in two basic operational designs—counterflow and crossflow—it might be helpful at this time to list a few terms and their definitions to aid in investigating these differences. The letters in the circles on the schematic diagrams in Figures 3-17 and 3-18 refer to the elements that perform the same function and their location in the structure. The giant natural draft or hyperbolic cooling towers operate on the same principles as a small 50-ton tin-box package; only the size is vastly different.

UPGRADEABLE ELEMENTS

To upgrade the performance of an existing cooling tower, the three major areas to investigate for retro-fit are:

1. Wet decking fill
2. Water distribution system
3. Drift eliminators

WET DECKING FILL

Generally, the most significant improvements can be made simply by changing the wet decking fill. This, however, is not done capriciously. The heat transfer must be investigated from a thermal engineering point of view, in conjunction with the fill characteristics determined by the manufacturer's performance curves. These are developed very painstakingly by trial, error and experimentation, which is expressed as KaV/L* or heat transfer characteristics.

The basic function of a cooling tower is to cool water by intimately mixing it with air. This cooling is accomplished by a combination of sensible heat transfer between the air and the water and the evaporation of a small portion of the water. This type of transfer is represented by the Merkel equation:

$$^*\frac{KaV}{L} = \int_{T_2}^{T_1} \frac{dT}{hw-ha}$$

NOMENCLATURE

Where:

a   area of transfer surface per unit of tower volume; expressed in square feet per cubic feet

dT   (delta T) difference between hot and cold temperature expressed in °F

h_a enthalpy of air-water vapor mixture at wet bulb temperature; expressed in BTU per pound of dry air

h_w enthalpy of air-water vapor mixture at bulk water temperature; expressed in BTU per pound of dry air

K   overall enthalpy transfer coefficient; expressed in pound per hour per square feet per pound of water per pound of dry air

L   mass water flow; expressed in pound per hour per square feet of plan area

T_1 hot water temperature; expressed in °F

T_2 cold water temperature; expressed in °F

V   effective cooling tower volume; expressed in cubic feet per square feet plan area

Figure 3-19.

Wood splash pattern. Observe the big distances between splash bars.

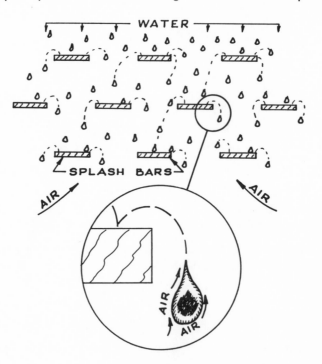

Figure 3-20.

Schematic operation of splash bar cooling.

It is quite obvious that this wood fill (Figure 3-19), is extremely inefficient compared to the cellular fill; plus the fact that the cellular fill has a lower resistance to air flow, which further enhances the heat transfer by more efficient utilization of the existing air. In wood slat splash bars, (Figure 3-20), droplets of water bounce from one layer of wood to the other and the rising air cools the outside of each sphere of the water droplets. Cellular fill takes the same droplets of water and spreads it out in a very thin molecular film where the air can now affect the entire surface of the film (Figure 3-21). Considerably more surface is then available to the flowing air for vaporization and sensible heat exchange to take place. The film pack contains more surface area than splash bars, and since the design of cellular fill permits air to flow through it with less resistance, it is extremely efficient compared to the old-fashioned splash bar mixture system for the same parameters of cooling.

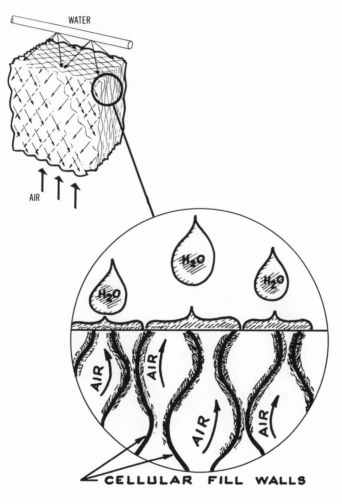

Figure 3-21.

Schematic illustration of cellular film fill cooling.

Crossflow cooling towers also have a rebuilding capability to improve the performance by changing the wood splash bars to more efficient fill. Old-fashioned stacked fill is rather inefficient compared to the new grid type splash bar packing consisting of corrosion proof fiberglass polyester grids for the higher capacity installation. A less expensive approach is to use metal grid hangers that are coated for corrosion protection. The installation of this type of fill is quite laborious, but is relatively inexpensive to install for the percentage of improvement that can be obtained compared to stacked fill.

The greatest improvement of performance in crossflow fill is obtained by changing the wood splash bars to cellular fill, (Figure 3-22), installing redistribution sections of predetermined levels, and adjusting the flow rate throughout the hot water distribution basin to provide proper water loading through the fill.

Figure 3-22.

Crossflow tower converted to cellular fill.

Here again, wood splash bars are replaced by the efficient cellular fill sections, thereby reducing air flow resistance through the tower. This provides a more effective use of the input energy, and results in colder water from the tower. Whether it be crossflow or counterflow, *uniformity is the key to success:* uniformity of water distribution, uniformity of fill configuration, and uniformity of air velocity—even though in a crossflow tower, the air pressure differential will vary with the height.

WATER DISTRIBUTION SYSTEM

For maximum performance, the water distribution system must provide a uniform pattern over the fill for optimum air-water interface. The older type trough, (Figure 3-23), has a very uneven splash distribution based upon columns of water falling vertically and hitting cups that, when accurately placed, distribute the water throughout the fill. Over the years this delicate balance is destroyed as the tower deteriorates. The end result is a vertical column of water, in many places, dropping up to four feet through the fill before it is broken up, losing entire areas of efficiency. When a nozzle is clogged, it leaves a dry spot on the fill; the air, being lazy, follows the path of least resistance and rushes up this dry spot, wasting energy and cooling potential.

Figure 3-23.

By replacing this water trough with ceramic spray nozzles and PVC piping, tremendous upgrading in capacity is available.

If a given cooling tower is performing well, its pipes are not corroding, the nozzles are not clogged up, and the maintenance costs are negligible, then there is no requirement to spend money on the installation. On the other hand—if better performance through colder water is urgent, the maintenance department spends too many man-hours cleaning nonfunctioning nozzles (Figure 3-24), and management is considering spending capital investment money for new facilities, now is the time to consider upgrading the existing plant by changing the water distribution to a highly efficient method—utilizing cheaper tax deductible maintenance dollars.

Figure 3-24.

Package tower with 388 small 1/8″ diameter orifice spray nozzles which tend to clog up quite easily.

Experience and many installations have proven the high efficiency of water distribution of the specially designed ceramic nozzle, together with noncorroding polyvinylchloride (PVC) piping in Figure 3-25. Contrasted to Figure 3-24, which was designed for the same GPM of water, it is obvious that the labor savings in not having to clean the 12 ceramic 1″ orifice units periodically will more than offset the installation cost against the old-fashioned ⅛″ openings on the 385 nozzles.

Also the large diameter 3″ PVC piping will not clog with rust readily. If by chance it does, it is simple to unscrew the end cap for blowdown. The nozzle, (Figure 3-26), snaps out of a bayonet slot holder by hand pressure.

The above portion has been devoted to small package units; however, the same techniques apply to larger field-erected installations.

The necessary hydraulic calculations will determine the pipe sizes and orifice openings. Figure 3-27 shows a 14″ diameter pipe assembly with 6″ diameter arms reduced to 4″ diameter, with the nozzles on 42″ centers, 42″ between arms. The

Figure 3-25.

Same tower with 12″-wide orifice (1¼″ diameter) spray nozzles handling the same gallonage as Figure 3-24.

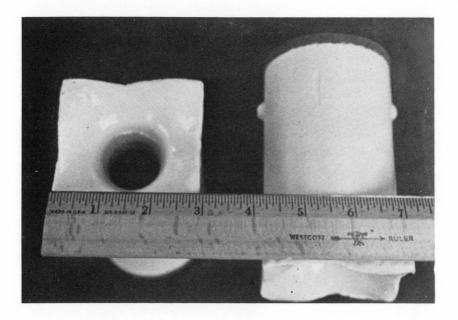

Figure 3-26.

Nonclogging, noncorroding, large-diameter ceramic nozzle produces a square spray pattern.

water distribution pattern on Figure 3-28 indicates a slight overlap of the cone in the previous installation. This depth is calculated and located at the optimum point by the manufacturer's curves and charts.

If existing steel header pipes and arms are corroded to the point of replacement, the labor cost of installing PVC piping over galvanized steel is less. A man with a

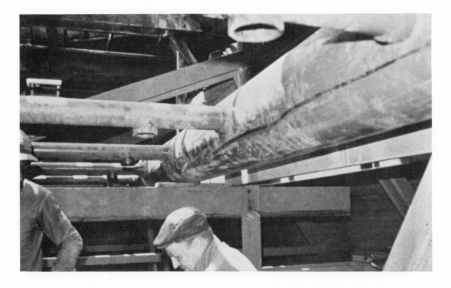

Figure 3-27.

14″ diameter PVC main header with 6″ diameter branch arms having nozzles on 42″ center distances.

Figure 3-28.

Spray pattern from ceramic nozzle provides complete wetting coverage of fill.

rope can hoist the plastic pipe and have an assistant help lock it in place. The weight of the steel will require riggers and a large team of men to do the same thing. (See Figure 3-29.)

Figure 3-29.

Ease of installation of hauling up 12″ diameter by 12′ long PVC piping.

Water volumes from 15 GPM to 250 GPM per nozzle are available in the two to seven psi-pressure range, well within the operating parameters of today's cooling tower requirements.

APPLICATION OF MODERN TECHNOLOGY

Using the previous information about the various elements, let us investigate examples of the application of modern technology to permit the operation to produce additional profits after a payback of the retro-fit costs.

1. Industrial Plant Refrigeration Systems

At a food processing plant on the East Coast, the total water on the tower is 3,750 GPM (14,200 liters per minute) for two cells. Design conditions were to cool the entering water from 95°F (35°C) leaving at 85°F (29.5°C) at an ambient wet bulb temperature of 78°F (25.5°C). One cell at a time was converted, (Figure 3-30), to maintain continuous plant operation.

The first indication that engineering calculations and predictions would be exceeded occurred when the total amount of gallons was put on the newly converted cell, while the second cell was rebuilt. With 50 percent of the tower in

Figure 3-30.

Conversion of one cell indicated that it could cool entire water volume to design. Completion of second cell reduced the cooling tower return water by 4°F.

operation and the same electrical input to the 40 HP motor, the one cell cooled the water to design with a 10°F (5.5°C) range and 7°F (4°C) approach. When both cells were operated with maximum water 3,750 GPM (14,000 liters), temperatures were returned almost 4°F (2.3°C) colder.

Records kept by the chief engineer indicated there was a direct fuel oil savings of approximately $300 a day, since in generating their own steam for the turbines, the 6,000 gallons-a-day (22,700 liters) oil requirement was cut down to 5,700 gallons (21,500 liters)—a savings of 300 gallons (1,200 liters) a day.

Further, the turbines were running at 350 GPM less and head pressures were four to five pounds per square inch (0.28 to 0.35 kg. per square cm) lower than previous records indicated. Those small increments added up to approximately $60,000 a year savings. With escalation of all sorts prevalent, this figure should also increase.

The ultimate savings, however, is that there is a 50 percent plant expansion contemplated. Calculations indicate that, with the addition of 12″ (30 centimeters) of cellular fill and piping changes to larger diameters for the water distribution system, the existing cooling tower with two 40 horsepower motors and 16′ (4.9 meters) diameter fans would be more than adequate to take care of the new requirements. This should result in an economy of approximately $80,000, which is the cost between new OEM equipment and additional rebuilding work.

2. Office Building Air Conditioning

This involves a counterflow tower in a large government office bulding in Washington, D.C., where original design conditions were marginally met. G.S.A. required a 25 percent increase in performance from 12,000 to 15,000 GPM (45,420 to 56,770 liters) per minute due to higher heat loads. The successful bid was $125,000 less than a proposal to install new original equipment manufacturers towers. Since the cooling tower was built using the vertical building columns as the main tower supports, cost of demolition and installation, including reconstructing the building around the new equipment, dictated that the prudent way would be to rebuild the existing tower.

One of the keys to improvement is a ceramic nozzle costing $10. This replaces a dozen or more bronze units $6 to $8 each, plus the cost reduction for using PVC piping at $14 per foot installed for 12″ diameter (30 centimeters diameter), versus $22 installed for galvanized steel. The weight differential is 90 percent plus in saving labor during installation.

The elements of rebuilding this tower are illustrated in Figure 3-31.

Figure 3-31.

Elements of Upgraded Rebuilt Tower:
a. Cellular Fill
b. Drift Eliminators
c. PVC Piping
d. Ceramic Nozzles

A. 30 inches (76 centimeters) of cellular fill replaced 144 inches (365 centimeters) of wood splash bars.

B. Efficient fireproof drift eliminators that can be walked on, provided mist protection and also acted as a fireproof barrier.

C. A tremendous aid in upgrading this structure was the replacement of old style gravity trough water distribution to PVC low pressure piping.

D. In conjunction with the PVC piping, ceramic nonclogging, noncorroding large orifice nozzles were installed.

This type of rebuilding is an open-ended funnel. Additional capacity may be obtained by pumping more water over the fill, increasing the depth of fill, or increasing the airflow; any combination of these events will add up to more work for less investment. In this case, the savings in dollars was generated because the four 60 horsepower motors, after conversion, produced an additional 3,000 GPM of cooling (11,330 liters per minute). If the four-cell tower was not rebuilt and an additional cell installed, then 300 horsepower would have been required to produce the same amount of work that the presently rebuilt installation is doing using only 240 horsepower. Multiplying the current Washington, D.C. power rates by the tower utilization, results in a KWH reduction of 1,575 and a $11,814.15 savings per year.

3. The Interchurch Center

The design conditions of this four-cell, double flow crossflow tower were originally to cool 5,400 GPM from entering the tower at 95°F leaving at 85°F at a 78°F wet bulb temperature.

According to the manager, design conditions were never met and the deficiency of the cooling tower created high heads requiring excess electricity to power the system.

Since the Interchurch Center is not a profit-oriented organization, funds were limited. Thermal engineering calculations indicated that if two of four cells were rebuilt with highly efficient cellular fill, the results would bring the tower up to design.

All four cells were stripped of wood and the interior metallic surfaces were sandblasted to remove corrosion and the old coating, primed with synthetic Parlon®, and coated with two applications of Corro-tect®, liquid chlorinated rubber corrosion protective coating.

The wood splash bars were carefully separated so that sufficient quantities were selected to rebuild two cells. The remaining two cells were then installed with cellular fill.

Operating records indicate that the condenser water is now two degrees colder than last year, which is more than adequate to provide proper operating conditions. The head pressures are running five to six psi lower, which reduces electrical consumption. One of the major savings is that in their two-refrigeration machine operation, one machine is now utilized 25 percent less than previous seasons. Future planning calls for changing the remaining two wood cells. A further improvement will produce extra colder water, which will more than pay back in energy conservation, and includes projected less maintenance wear and tear on the equipment.

4. Sage Intercontinental Defense Missile Command

This cooling tower was 18 years old and required extensive rebuilding as an alternate to demolition or replacement. Government specifications called for rebuilding the existing drift eliminators, taking out and putting back new wood wet decking fill, installing gravity splash nozzles attached to the wood water distribution trough orifices, miscellaneous carpentry work, and repairs.

As an alternate bid, the U.S. Air Force engineers, after an educational session, re-advertised the bid, allowing an alternate for modern retro-fit. Drift eliminators fabricated from cellular materials were installed, water troughs were taken out, and PVC piping with ceramic square spray nozzles were engineered and installed together with wet decking cellular fill.

This new system saved the Air Force $25,300 over the second bidder, and upon testing the installation with full load, the cooling tower performed 32 percent over the previous year's operation of cooling 5,000 GPM from entering the tower at 95°F, leaving at 85°F at 75°F ambient wet bulb temperature.

One of the main upgrading elements was changing the old water trough distribution system, (Figure 3-23), to PVC piping and ceramic nozzles (Figure 3-28).

CONCLUSION

A blanket statement cannot and should not be made that all cooling towers can be rebuilt to save large quantities of money, power or can return water colder by a significant amount. Each installation must be treated on an individual basis with thermal, hydraulic, and aerodynamic calculations studies of existing conditions to see where areas of improvement can be largely anticipated. Since significant sums of money can be spent on larger cooling towers, it is strongly recommended that a testing requirement be made part of the contract. The Cooling Tower Institute (CTI) has formulated ATC-105, which contains rigid testing procedures for determining whether or not a cooling tower will produce what the customer has been promised and is expected to pay for.

Acceptance Test Procedure Code is recognized throughout the industry as the standard measurement of a tower's efficiency. The CTI, in the interests of obtaining a strictly impartial test, contracts the testing to be done by the Midwest Research Institute (MRI), whose only concern is to conduct the procedure according to the code, collect field data, using authorized and calibrated instruments, and obtain a computerized readout of the percent of the cooling tower's performance.

All interested parties, the contractor, the owner, and the manufacturer, are invited to observe to see that impartiality is conducted; and regardless of who pays the fee, the owner receives a report as to the operation of his cooling tower.

A cursory inspection by maintenance managers or consulting engineers of a deficient and seemingly old cooling tower should not automatically bring forth the snap judgment that this structure should be removed and a new unit put in. Examination of the tower's structural strength and its components could indicate that, with the applications of sound modern engineering principles and proper maintenance, the cooling tower could be given an extended useful life of service, producing colder water for the added profit of the facility and saving input energy.

4

Air Compressors and the Compressed Air Systems

William Scales, P.E.

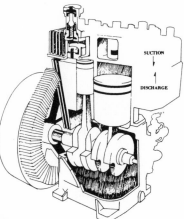

SCOPE OF CHAPTER

Many problems with air compressors and the compressed air system can be avoided with proper selection of the equipment. This chapter deals with selection, applications, and evaluations, and provides details on what and how to do maintenance to prolong the life of air compressors and accessories.

Ask most people about applications of air compressors and perhaps one or two can think of an answer. Most people are unaware of the diversity of compressed air—that it affects their everyday lives from the aeration of the drinking water in a reservoir, to the making of artificial snow, to the operation of air tools, spray equipment, and packaging machinery in manufacturing plants throughout the world.

There are two basic types of air compressors in use today. The positive displacement type compresses air by admitting successive volumes of air into a closed space and then decreasing the volume. Reciprocating, rotary screw, sliding vane, two lobe rotary, and liquid piston are all examples of a positive displacement compressor. Dynamic compressors refer to the centrifugal or axial flow compressors that apply mechanical force to the air to increase its velocity, which is then converted to pressure. The majority of industrial plants have positive displacement compressors; therefore, the major portion of this text is devoted to this type, although the same principles apply to maintenance. The sections dealing with the compressed air system and accessories are applicable to any plant regardless of the type of compressor used.

BASIC DEFINITIONS

Before one can discuss or select the compressor which is best-suited for an application, there are certain basic definitions that should be understood.

The pressure in a system is defined in psia (pounds per square inch absolute) and psig (pounds per square inch gauge). The absolute pressure is the sum of the gauge pressure and the atmospheric pressure. The basic metric unit used is kilograms per square centimeter; more recently the kilopascal has been accepted as the unit of measure of air pressure.

Volume is expressed in cubic feet per minute (cfm) in the English system, and in cubic meters per hour or per minute in the metric system. However, because there are modifications to the normal term *volume*, one must clearly understand whether reference is being made to actual cfm (acfm), actual cfm at inlet conditions, standard cubic feet per minute (scfm), free air (cfm), etc.

Brake horsepower (BHP) is the measured horsepower delivered to or required at the compressor shaft. One of the most commonly used measurements of a compressor's efficiency is the brake horsepower required to produce 100 cubic feet (BHP/100). The lower the figure, the more efficient the compressor. This figure enables the plant engineer to compute energy costs. At 100 psig, the BHP/100 can vary between 18.5 and 27 depending on the type of compressor selected. In many cases, the annual cost of electric energy exceeds the initial price of the air compressor. Therefore, careful selection of the compressor coupled with accurate application information can save a company thousands of dollars per year in energy costs.

Become familiar with the terminology used in the industry to be able to compare performances. (See Figure 4-1.)

COMPRESSOR TERMINOLOGY

ABSOLUTE PRESSURE (psia) is the total pressure measured from absolute zero, i.e., from an absolute vacuum. It equals the sum of the gauge pressure and the atmospheric pressure corresponding to the barometer. At sea level atmospheric pressure is 14.7 psia and 100 pounds gauge pressure would be 114.7 psia.

ACTUAL CAPACITY (acfm) is the actual volume rate of flow of air compressed and delivered at the discharged point but referred to conditions of total pressure, temperature, and humidity at the inlet.

ADIABATIC COMPRESSION is achieved by preventing the transfer of heat to or from the air during compression.

AFTERCOOLERS remove heat from the air after compression is completed and are used to effect moisture removal.

AIR RECEIVERS or "tanks" provide a volume of stored air. They are used as pulsation dampeners on reciprocating compressors and are generally sized in cubic feet for a minimum of one-seventh the cfm of the compressor for constant speed operation and one-third for automatic start-stop control.

BLOWERS generally refer to the broad category of rotating compressors which operate below 40 psig.

BOOSTER COMPRESSORS receive air that has already been compressed and deliver it at a higher pressure.

BRAKE HORSEPOWER (BHP) is the measured horse power delivered to the compressor shaft.

CAPACITY is often used interchangeably with actual capacity or inlet capacity and is stated in cfm.

CENTRIFUGAL COMPRESSORS are dynamic type compressors where velocity is translated to pressure, and discharge is through a collector ring at the outside rim of the housing. It is often built as a multi-stage unit.

CLEARANCE is the volume inside a compression space that contains air trapped at the end of the compression cycle—it is between the piston and the head and under the valves in a reciprocating compressor.

Figure 4-1.

COMPRESSION EFFICIENCY (adiabatic) is the ratio of the theoretical horsepower to the horsepower imparted to the air actually delivered by the compressor. The power imparted to the air is brake horsepower minus mechanical losses.

COMPRESSION RATIO is the ratio of the absolute discharge pressure to the absolute inlet pressure.

DISCHARGE PRESSURE is the absolute total pressure at the compressor's discharge flange; it is generally stated in gauge pressure (psig) but should include reference to the barometic pressure.

DISCHARGE TEMPERATURE is the total temperature at the discharge flange of the compressor.

DISPLACEMENT of a compressor is the volume expressed in cubic feet per minute (cfm) displaced per unit of time. The term is generally used for reciprocating compressors and is derived by taking the product of the net area of the compressor piston times the length of stroke and the number of compression strokes per minute. The displacement of a multi-stage compressor is the displacement of the low pressure cylinder only.

DOUBLE-ACTING COMPRESSORS compress air on both strokes of the piston, therefore twice per revolution.

DRYERS are machines or devices used to remove water vapor from the compressed air. The most common types are deliquescent, refrigerated and regenerative.

DYNAMIC COMPRESSORS refer to the centrifugal or axial flow compressors that translate velocity into pressure, i.e., kinetic energy to pressure energy.

FREE AIR is defined as air at atmospheric conditions at any specific location. Because the altitude, barometer and temperature may vary at different localities and at different times, it follows that this term does not mean air under identical or standard conditions.

INLET PRESSURE is the absolute total pressure existing at the intake flange of a compressor.

INLET TEMPERATURE is the total temperature at the intake flange of a compressor.

INTERCOOLING is the removal of heat from the air between stages. Ideal intercooling exists when the temperature of the air leaving the intercooler equals the temperature of the air at the intake of the first stage.

ISOTHERMAL COMPRESSION is achieved by maintaining the air at constant temperatures during compression.

LIQUID PISTON COMPRESSORS displace the air in an elliptical casing with water or other liquid, which acts as the "piston" to compress the air.

LOAD FACTOR is the ratio of the average compressor load during a given period of time to the maximum rated load of the compressor.

MECHANICAL EFFICIENCY is the ratio of the horsepower imparted to the air to the brake horsepower expressed in percent.

MOISTURE SEPARATORS are small vessels that are usually used with intercoolers and aftercoolers to separate the condensed moisture from the air through mechanical or cyclonic action.

MOISTURE TRAPS automatically discharge moisture from separators or receivers. There are many types available such as float, inverted bucket, solenoid valve with timer, etc.

MULTI-STAGE COMPRESSORS achieve final discharge pressure in more than one stage.

OVERALL EFFICIENCY is the ratio of the theoretical horsepower to the brake horsepower. It is equal to the product of compression efficiency times mechanical efficiency. Electric motor and transmission losses should also be considered in computing cfm out versus kilowatts in.

PACKAGED COMPRESSORS are complete units consisting of all components, i.e., compressor, motor, aftercooler, receiver and controls mounted on one base for "single point" electrical and air and two point water connections. Installation is simplified with this type of purchase.

PORTABLE COMPRESSORS are usually mounted on skids or wheels to be moved from one location to another. They may be electric, gas or diesel driven.

POSITIVE DISPLACEMENT compressors compress air by increasing pressure by admitting successive volumes of air into a closed space and then decreasing the volume. Reciprocating, rotary screw, sliding vane, two lobe rotary, and liquid piston are all examples of a positive displacement compressor.

RECIPROCATING COMPRESSORS achieve compression by a piston moving in an enclosed cylinder and decreasing the volume.

ROTARY SCREW COMPRESSORS utilize two close clearance helical lobe rotors turning in synchronous mesh and compress the air by forcing it into a decreasing inter lobe cavity until it reaches the discharge part.

SINGLE-ACTING COMPRESSORS compress air on only one side of the piston. For example, in vertical configuration the downward stroke is suction, and the upward stroke is compression.

SINGLE-STAGE COMPRESSORS accomplish final discharge pressure in one step.

STANDARD AIR (scfm) is air at defined conditions. Two standards are presently in use. The first is a temperature of 68 degrees F, a pressure of 14.7 psia and a relative humidity of 36 percent. The second is 60 degrees F, 14.7 psia and dry.

VACUUM PUMPS compress air from sub-atmospheric pressures to atmospheric.

VANE TYPE COMPRESSORS have a rotor and vanes or blades that slide radially in an offset housing forming sealed sectors. Volume is decreased by the convergence of the cylinder walls with the vanes and the rotor body.

VOLUMETRIC EFFICIENCY is the ratio of the capacity of the compressor to the displacement expressed in percent. It is usually between 60 and 80 percent and is always reduced as pressure increases in a given compressor.

Figure 4-1 (continued).

TYPES OF COMPRESSORS
(See Figure 4-2.)

The types of compressors are positive displacement and dynamic. The reciprocating, rotary sliding vane and rotary screw are examples of positive displacement machines and the centrifugal and axial flow are dynamic compressors. A description of each and an illustration appear in Figures 4-2 through 4-6.

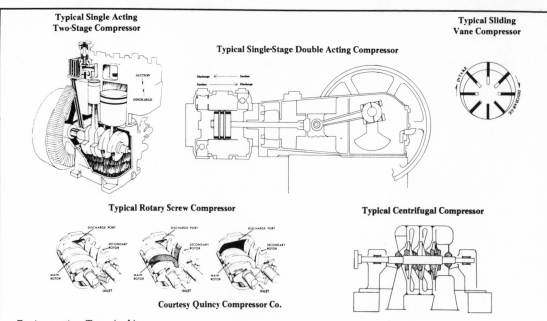

Typical Single Acting
Two-Stage Compressor

Typical Sliding
Vane Compressor

Typical Single-Stage Double Acting Compressor

Typical Rotary Screw Compressor

Typical Centrifugal Compressor

Courtesy Quincy Compressor Co.

Reciprocating Type (a, b)

Compression achieved by a piston moving in an enclosed cylinder and developing pressure is a reciprocating compressor. Cylinder arrangement may be vertical, horizontal or angular, and may be either air-cooled, water-cooled, lubricated or non-lubricated. Compression may take place in either one end (single-acting) or both ends (double-acting).

Single-stage compressors consist of one or more cylinders taking air in from the atmosphere and compressing it to the final discharge pressure required. Normally single-stage units are used for pressures up to 100 psig for continuous service and can sometimes be used for intermittent service at higher pressures.

Two-stage compressors have a minimum of two pistons; the first compresses air from atmosphere to an intermediate pressure (25-40 psig). The air is then cooled in an intercooler before entering the second stage where it is compressed from the inter-stage pressure to the final discharge pressure (100-200 psig). Due to the dissipation of heat between stages (because of the intercooler), above 100 psig, two-stage units allow for more efficient and cooler operating compressors, which increases compressor life.

Rotary Sliding Vane (c)

A rotary sliding vane compressor is a machine with a rotor and metallic or non-metallic vanes which slide radially in an offset housing forming sealed sectors. As the rotor turns, the vanes are held to the cylinder walls by centrifugal force. As a sector rotates toward the discharge port, its volume is decreased by the convergence of the cylinder walls with the vanes and rotor body. Units can be single or two-stage, water or air-cooled.

Rotary Screw (d)

Two close clearance helical-lobe rotors turn in synchronous mesh. Normal is four lobe main male rotor with a six groove mating female rotor (gate). Air is drawn into the suction port into a space between the rotors. As the rotors revolve, air is forced into decreasing inter-lobe cavity until it reaches the discharge port. In lubricated units the male rotor drives the female and oil is injected into the cylinder serving as a lubricant, coolant and as an oil seal to reduce back slippage. On non-lubricated types, timing gears are used to drive the rotors and multi-staging is necessary for pressures above 50 psig.

Centrifugal Compressor (e)

A compressor in which air is introduced at the center of a rotating impeller which then accelerates the air radially. This velocity is translated to pressure, and discharge is through a collector ring at the outside rim of the housing. It is often built as a multi-stage unit.

Axial Flow Compressor

A compressor in which the air is introduced at one end of the unit and accelerated along the axis of the compressor. This velocity is translated to pressure, and discharge is from the end opposite the intake. It is often built as a multi-stage unit.

Figure 4-2.

a. Typical Single-Acting Two-Stage Compressor
b. Typical Single-Stage Double-Acting Compressor
c. Typical Sliding Vane Compressor
d. Typical Rotary Screw Compressor
e. Typical Centrifugal Compressor

Reciprocating Compressors

In considering the purchase of a reciprocating compressor, a selection of single-acting, double-acting, single-stage, two-stage, air-cooled, and water-cooled units requires evaluation (Figures 4-3 through 4-7). While each plant is different, the plant engineer must weigh initial cost, installation, maintenance, and operating expenses to arrive at a fair comparison. In addition, if water is required for a water-cooled unit, or fresh air ducts are needed to keep the ambient temperature for an air-cooled unit below 100°F, then these costs must also be calculated. At 100 psig, multi-stage units generally result in a power savings of up to 15 percent versus single-stage units. In addition, multi-staging with cooling of the air between stages reduces the final discharge temperature in the cylinders, which reduces the problems with carbon deposits and lubrication breakdown. Generally, the higher the discharge temperature, the higher the maintenance and operating costs of a compressor, and the greater the possibility of discharge line explosions in lubricated compressors.

The types of controls that are available to regulate the delivery of the reciprocating compressor fall into two basic categories.

A. Automatic Start-Stop

In this form of control, a pressure switch functions to control automatically the starting and stopping of the compressor. This type of control is generally used on smaller compressors (tank mounted type) or on larger compressors when the demand for air is quite low (less than 50 percent). It is important when using start-stop control that the air receiver be sized to limit the number of starts per hour to a figure that is acceptable to the motor manufacturer. Frequent starts can reduce motor life and create control problems and rapid contact wear in motor starters.

B. Constant Speed Control

In this mode, the compressor operates continuously at its normal operating speed and the compressor is "unloaded" (compressor stops pumping but continues running) at a set point. When pressure drops to a preadjusted point, the unloading mechanism allows the compressor to compress air again. The primary method of unloading is inlet valve regulation. This system uses a device (air-operated pilot valve or a pressure switch with a three-way electric solenoid valve), which allows air to feed back from the receiver to a plate actuated by a diaphragm or unloading piston on top of the inlet valve. This plate holds the intake valves open whenever there is no demand for air (Figure 4-8). With the inlet (also called intake or suction) valves open, the air will sweep in and out of the cylinder and very little horsepower is required to operate the compressor. Generally, the combination of the friction horsepower and horsepower required in the air to force it through the inlet valve is less than 20 percent of the full load horsepower.

Figure 4-3.

Air-cooled tank-mounted compressor available through 25HP tank-mounted and through 125HP base-mounted.

Figure 4-4.

Tank-mounted, single-acting, water-cooled, two-stage compressor. Available through 150HP base-mounted.

Courtesy of Joy Manufacturing Company

Figure 4-5.

Vertical single-stage, double-acting, water-cooled compressor. Available in lubricated cylinder or in teflon ring construction through 125HP.

On smaller compressors loading and unloading is normally accomplished in two steps, i.e., either fully loaded or full unloaded. On larger water- cooled compressors, three-step control (full load, half load, no load) and five-step control (100 percent, 75 percent, 50 percent, 25 percent, no load) is available.

Dual control combines the automatic start-stop and constant speed control so that the operator can select the mode required based on the plant demands. For example, constant speed operation may be required during the normal workday and start-stop may be used at night and on weekends when demand for air may be lower.

Automatic dual control is a further refinement. This system has constant speed control as the basic mode of operation. If the compressor unloads and remains

Courtesy of Chicago Pneumatic

Figure 4-6.

Horizontal single-stage, double-acting, packaged, water-cooled compressor. Available in lubricated or in teflon ring construction through 125HP.

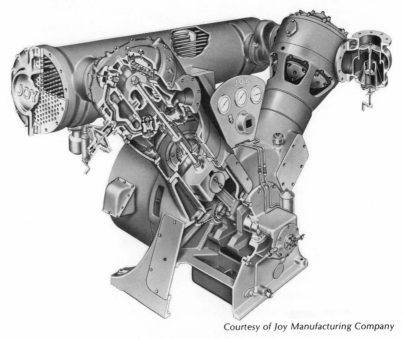

Courtesy of Joy Manufacturing Company

Figure 4-7.

Two-stage, double-acting, water-cooled compressor. Available in lubricated or in teflon ring construction through 400HP. Different configurations available in larger sizes.

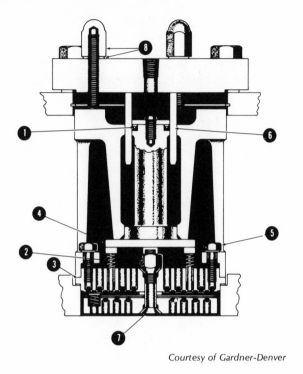

Figure 4-8. Suction Unloading Valve Assembly:

1. The plunger seal is used to seal the unloading air.
2. Finger pin springs return the unloader mechanism when unloading air is released.
3. Clamp rests on full circumference of valve seat to eliminate warpage.
4. Plunger acts on finger pins to open valve disc fully and evenly.
5. Clamp is secured to the valve assembly by hex head cap screws to aid in installation and removal.
6. Separate valve cover is guided in clamp to provide proper alignment of assembly.
7. Complete lower valve assembly is held together by one flat head screw and hex nut permitting easy assembly and disassembly. Numerous suction and discharge valve parts are interchangeable.
8. Acorn nuts and gaskets prevent leakage around locking screws.

unloaded for a preset period of time, the compressor stops and is placed in the standby position. If pressure continues to drop, the compressor will start automatically and begin the cycle again.

Dual and automatic dual control are very inexpensive compared to the purchase price of a larger compressor, and these options should be requested as part of the quotation when specifying or purchasing new compressors.

SAFETY DEVICES

The number of devices used in a compressor system depends upon the plant engineer and the compressor vendor. However, the most commonly used are:

A. High Air Temperature Switch

This is installed in the air compressor discharge line as close to the discharge flange as practical. At 100 psig, the switch is set approximately 400-425°F on a water-cooled single-stage compressor and 300-325°F on a two-stage compressor. This switch is very important and will normally shut down the compressor because of inadequate cooling or bad valves. On air-cooled machines, the settings may be somewhat higher.

B. Low Oil Level and Pressure Switches

1. Low oil level switch

This monitors oil level on a "splash" lubricated unit and will not allow the compressor to start unless adequate oil is in the crankcase.

2. Low oil pressure switch

This monitors oil pressure on a pressure lubricated compressor. On an oil pressure switch, the contact is normally open and closed on pressure rise. Since this contact is in series with the coil of the magnetic starter, a normally closed contact, which is timed to open after 10 or 15 seconds, (enough time for oil pressure to develop), is shunted across the oil pressure switch contact to allow the compressor to start. If oil pressure has not developed, the compressor will shut down and should be locked out through a proper relay.

C. Safety Valves

Safety valves should be used:

1. on all receivers.
2. before any isolation valves that may be installed between the compressor head and the receiver.

Safety valves should be of the ASME type and must be set at least 5 psi below the working pressure of the receiver it is protecting. The safety valve should also be used between the compressor and any isolation valve between the compressor head and the receiver. This should be set at the maximum safe operating pressure of the compressor or below the maximum safe operating pressure of the isolation valve. A safety valve on an interstage cooler will not protect the system against the danger of an unprotected isolation valve. The interstage valve merely protects the low pressure cylinder from damage that may be caused when a valve malfunction occurs on the high pressure cylinder. Safety valves are also often installed before aftercoolers to protect against heavy carbon buildup restricting the flow of air through the cooler. Carbon restriction could result in excessive head pressure and possible damage to the compressor or fire and explosions in the discharge line.

VALVES

Valves in a compressor operating at 900 rpm for only eight hours per day and 220 days per year open and close 54,000 times per hour or 95,000,000 times per year.

The valves are the heart of the compressor and must be maintained properly for the compressor to operate properly. Whether valves are of concentric ring design or rectangular leaf or channel type, they should be inspected yearly as a minimum and rebuilt as experience dictates. Keeping records of valve repairs and what has worn is of primary importance in determiing causes of failure. Replacement parts should be kept in stock to reduce downtime that may result because of a valve failure. The parts should be greased if necessary and wrapped in a protective coating before storing in a dry area.

Follow the manufacturer's instructions carefully when rebuilding valves and use only genuine replacement parts. Well-designed valves should not require frequent repairs. However, poor repair work or foreign matter will increase the frequency of repairs. Dirty intake filters, excessive lubrication, or the wrong lubricant and moisture in the air all contribute to valve failures.

In water-cooled compressors, cylinder jacket water temperatures should be maintained 15-20°F above the incoming air temperature to insure that condensation of the air does not cause water (condensate) in the valves and cylinder areas. The water temperature can be maintained automatically with a thermostatic valve or monitored by plant personnel and adjusted manually as required. (Because of varying conditions, manual adjustment may not be practical.) The selection of the proper intake filter and lubricant will result in increased valve life.

Valve repairs require careful attention. Intake and discharge valve assemblies and parts may look alike but be different. It is important that these valves and parts be assembled *exactly* as they were removed.

Other important hints are:

1. Be sure the original valve lift is maintained.

2. Cleanliness of all valve parts is absolutely essential.

3. Inspect valves and springs carefully and replace if any wear is evident.

4. Replace valve seat gaskets and cover gaskets. In some cases these gaskets can be reused, but good practice dictates new gaskets.

5. Be certain inlet valves are placed in the inlet cavity and discharge valves in the discharge cavity. Severe problems can develop if the valves are inadvertently interchanged.

6. Valves can be checked for leakage after rebuilding by pouring a commercial solvent into the valve disc and observing if any liquid leaks past the valve disc(s). Be sure all solvent is removed before the valves are installed.

   ■ An often used method to test discharge valves is to install them in the cylinder or head without the suction valves and apply back pressure to allow the system to act on the discharge valves. If the valves leak, air will be in the cylinder and you will feel which valves leak.

7. Above all, read the instructions.

Clearances and Tolerances (This is intended as a guide <u>only</u>.)

Valves:

Lift is the primary dimension that is considered.
- ■ Concentric ring valves generally require .80 to .100 inches. However, one manufacturer states .060 to .070 inches for his valves.

- ■ Channel valves, feather valves and any other rectangular-shaped valves should be checked with the manufacturer's recommendations.

Rings:

Side clearance—.002 to .004 inches.
End gap clearance—.002 to .004 inches times piston diameter.

Piston to Cylinder:

Aluminum—.003 inches times the bore diameter plus .010 inches.
Cast iron—.001 inches times the bore diameter plus .005 inches.

Piston End Clearance:

Allow for expansion of the piston rod. From the piston face to the frame end should be one-third of the total and opposite piston face to the outer head should be two-thirds of the total clearance. This approximation should be done when the machine is cold. It can be done with a piece of lead or solder compressed by the piston against the head when the machine is rotated one revolution by hand. The lead or solder is inserted through a valve cavity.

- Packing side clearance:
  Cast iron—.005 to .010 inches.
  Teflon—.009 to .015 inches.

- Cross head clearance:
  .00075 to .001 per inch of diameter of the crosshead guide.

- Reboring cylinders:
  Reboring is practical in many cases where cylinders are out of round. The normal maximum reboring limits are .060 to .125 inches. Oversize pistons should be used in cases where boring is done.

  In many cases where cylinders are worn less than .010 to .020 inches, reboring is not necessary and oversize rings are all that is required.

TROUBLE-SHOOTING

Trouble-shooting can be accomplished effectively only if the operator has records of what normal operations should be. For example:

1. Record normal full load current at a specific voltage. These should be measured several times during the first week of operation of the compressor.
2. If the compressor has its own receiver, allow the compressor to fill this receiver from zero to the cutout point. Record the cutout pressure and the time required to fill the receiver.
   The above can be used as a guide to check compressor efficiency in the future.
3. Do not overlook an operator's instinct or feeling that something doesn't sound right. This knowledge of how a machine should sound has often prevented major damage.

MAINTENANCE

The best advice anyone can give is to read the instruction book. No manufacturer wants problems with his equipment and he knows his machinery better than anyone else. Follow his advice.

Basic maintenance and trouble-shooting instructions for both single-acting and double-acting reciprocating compressors follow. However, there is no substitute for the owner's instruction manual; at all times the owner's manual should be followed whenever there is any conflict with the guide in this text.

RECOMMENDED INSPECTION, MAINTENANCE AND SERVICE PROCEDURES FOR SINGLE ACTING AIR COMPRESSORS

DAILY

1. Check oil level. Replenish, if necessary, with proper oil and correct viscosity for surrounding conditions. Check oil pressure if unit is pressure lubricated.

2. Drain receiver of accumulated moisture. If automatic drain is used, manually drain tank weekly to check operation of automatic drain. Recommended that discharge of automatic drain be visible to check for operation.

The above can be done at bi-weekly intervals if operating conditions indicate this is adequate. Special attention should be given to changing seasons requiring more or less frequent attention.

MONTHLY

1. Check distribution system for leaks.
2. Manually operate safety valves to be certain they are functioning.
3. Clean intercooler fins and cylinder with a jet of air.
4. Replace or clean intake filter.
5. Inspect oil for contamination and change if necessary.
6. Check belts for correct tension.
7. Check operation of controls.
8. Check efficiency of compressor (pump up time check).

EVERY THREE MONTHS

1. Change oil.
2. Tighten all bolts.

YEARLY

1. Inspect valve assemblies. (May be necessary more frequently under heavy duty conditions)
2. Inspect motor starter, controls and wiring.
3. Check electric motor and lubricate if necessary.

TROUBLE SHOOTING

Fails to attain pressure 1, 2, 3, 4, 13, 14, 16
Knocking 2, 7, 11, 13, 15
Oil Pumping 2, 6, 9, 11, 13, 16
Compressor overheating 1, 2,
Motor draws excessive current 2, 5, 8, 10
Compressor doesn't unload when stopped 2, 3, 12
Abnormally high pressure 3,12,14
Compressor won't attain full operating speed 2, 3, 4, 10, 12
Excessive starting and stopping 1, 2, 14

1. Air leaks in distribution system.	9. Oil viscosity low.
2. Broken, leaking or carbonized valves.	10. Unbalanced or low voltage, loose connections or bad contacts.
3. Faulty unloader valve.	11. Worn bearings or connecting rods.
4. Loose belts.	12. Worn unloader parts.
5. Tight belts.	13. Piston, rings or cylinder worn.
6. Oil level high.	14. Faulty pressure switch.
7. Oil level low.	15. Loose flywheel, motor pulley or other loose parts.
8. Oil viscosity high.	16. Dirty intake filter.

The above is a general chart applicable for many compressors.

Figure 4-9.

Recommended inspection, maintenance and service procedures for single-acting air compressors.

DOUBLE-ACTING COMPRESSORS MAINTENANCE AND SERVICE

Daily

1. If pressure lubricated, check oil pressure.
2. Check oil level in compressor crankcase.
3. On lubricated units, fill the lubricator at least once per shift.
4. Observe cylinder lubricator sight feeds to be sure all feeds are operating at the proper feed to all cylinders and packing.
5. Check pressure controls for proper operation.
6. Drain moisture trap to pressure controls.
7. Drain water from bottom of receiver.
8. Check discharge temperature of cylinder cooling water.
9. If equipped with clearance pocket, drain moisture.
10. Check and investigate any unusual operation or noise.

Monthly

1. Drain oil filter and clean oil filter element.
2. Check and adjust (if necessary) all pressure and temperature shutdown devices.
3. Check and clean (if necessary) intake filter element.
4. Check all automatic moisture traps for proper operation.
5. Check operation of safety valves.
6. Check all pressure gauges for accuracy.

Semi-annually

1. Remove cylinder valves and inspect for wear, broken valve springs, broken valve discs, and damaged valve seats. If necessary, clean and repair.
2. Drain crankcase oil, clean interior of crankcase, and refill with new oil.
3. Inspect cylinder bore to be sure cylinder is receiving proper lubrication.
4. Check for loose foundation bolts and tighten if necessary.

Figure 4-10.

Double-Acting Compressors—Maintenance and Service.

5. Drain oil and clean force feed lubricator oil reservoir.

6. If compressor piston assembly is of the nonlubricated type, remove piston and check piston rings, piston, piston rod and compressor cylinder bore for wear.

7. Remove piston rod packing and piston rod oil scraper ring. Check for wear and clean (if necessary). *Important:* Don't intermix packing ring or oil scraper ring segment.

Annually

1. Remove, clean and inspect intercooler tube bundles, aftercooler tube bundles and clean interior of cooler shells.

2. Drain water and clean cylinder and cylinder water jackets.

3. Check and clean (if necessary) compressor motor.

4. Check and inspect compressor drive belts for wear and adjustment.

5. Check for wear and inspect main bearing, crank bearing, and wrist pin bushings.

6. Check crankshaft counterweights for tightness.

7. The same procedure should be followed for annual inspection as described in daily, monthly and semi-annual inspection information.

Above are intended as a guide only. Refer to manufacturer's instruction book for more specific information.

Figure 4-10 *(continued).*

LUBRICATION

A good lubricant should minimize wear, reduce friction, and minimize deposits. Provide the oil company with the specifications given by the compressor manufacturer and have the company agree that this oil is acceptable for the application.

The proper oil is probably neither the most expensive nor the least costly. Oil selection should never be based on price; it should be based on performance. If experience has resulted in low maintenance and carbon deposits, continue using the oil.

A single-acting lubricated compressor uses the same oil for the crankcase and the cylinder. While the function of the rings is to control and reduce oil carryover, some oil will pass. The oil in contact with the hot valve areas may leave carbon. A low carbon residue rating for the oil is recommended, although this doesn't

Double Acting Compressors

TROUBLE-SHOOTING CHART

NUMBERS IN SYMPTOM COLUMN INDICATE ORDER IN WHICH POSSIBLE CAUSES SHOULD BE TRACED

POSSIBLE CAUSES

	Failure To Deliver Air	Insufficient Capacity	Insufficient Pressure	Compressor Running Gear Over Heats	Compressor Cylinder Over Heats	Compressor Knocks	Compressor Vibrates	Excessive Intercooler Pressure	Intercooler Pressure Low	Receiver Pressure High	Discharge Air Temp. High	Cooling H$_2$O Discharge T High	Motor Fails To Start	Motor Over Heats	Valves Over Heat
Restricted Suction Line		4							4						
Dirty Air Filter		3							3						
Worn or Broken Valves L. P.	2	1	2		2				1		1	3	4	3	3
Worn or Broken Valves H. P.								1							
Defective Unloading System L. P.	1	2	1			7	4		2		1	2	4	3	4
Defective Unloading System H. P.								2							
Excessive System Leakage	5	3													
Speed Incorrect		6	6	3	3	7									
Worn Piston Rings L. P.	7	4							5						
Worn Piston Rings H. P.								3							
System Demand Exceeds Compressor Capacity		5													
Inadequate Cooling Water Quantity					4						4	1			
Excessive Discharge Pressure				4	1	11	9			2	3	5		2	1
Inadequate Cylinder Lubrication					6	10	8				8				
Inadequate Running Gear Lubrication				1		1								5	
Incorrect Electrical Characteristics													2		
Motor Too Small													5	1	
Excessive Belt Tension				2									7	7	
Voltage Low													6	6	
Loose Flywheel or Pulley						7	2								
Excessive Bearing Clearance						5									
Loose Piston Rod Nut						4									
Loose Motor Rotor or Shaft						9	6								
Excessive Crosshead Clearance						3									
Insufficient Head Clearance						2									
Loose Piston						6									
Running Unloaded Too Long															2*
Improper Foundation or Grouting						8	5								
Wedges Left Under Foundation							10								
Misalignment (Duplex Type)							3								
Piping Improperly Supported							1								
Abnormal Intercooler Pressure											7	7		4	
Dirty Intercooler											6	6			
Dirty Cylinder Jackets					5						5	2			
Motor Overload Relay Tripped													1		

* Inlet Valves

Figure 4-11.

Compressor Trouble-Shooting Chart.

necessarily guarantee low carbon deposits. If the compressor begins to pass oil, it is important that the problem be corrected rapidly. The danger of fires and explosions is present in all lubricated compressors using petroleum base lubricants. While the condition occurs infrequently, violent explosions can result where heavy carbon deposits have formed. One manufacturer has said that one ounce of oil per 50HP hours is acceptable while one ounce per 25HP indicates repairs are necessary. Of course, this is a rule of thumb and each machine is different.

"Proper" oils and viscosities depend on conditions. A suggested lubricant for single-acting machines would be a straight mineral oil with rust and oxidation inhibitors. For double-acting machines, a guide for cylinder lubricants would be as follows:

- Flash point—400°F
- Viscosity SSU at 100°F—245-420
- Viscosity SSU at 210°F—45-90
- Pour point—30° maximum
- Conradson carbon residue, percentage—1.0

In double-acting machines, the frame or crankcase is lubricated by constant circulation. This may require changing once every six months. Generally, the cylinder is fed from an external lubricator that is driven by a mechanical linkage. The lubricator is adjusted to supply the cylinder with the proper amount of oil. While this is measured in drops, it is an unprecise and unreliable measure. Valves and cylinders that are properly lubricated will have a light film of oil. Pools of oil or concentrations of oil in valve pockets indicate too much lubrication. Inadequate lubrication will be evidenced by dry cylinders and valves. Once the proper rate has been established, drops can be counted and used as the measurement—providing the oil characteristics remain constant and the lubricator is in good condition. If the oil consumption rate changes, inspect the lubricator for malfunctioning. A formula for estimating the amount of oil to inject into a cylinder is:

$$\frac{B \times S \times N \times 62.8}{10,000,000} = Q$$

B = bore (inches)

S = stroke (inches)

N = rpm

Q = quarts of oil per 24-hour day

Synthetic lubricants have been used more extensively in the past few years. The primary objection has been the cost and the possible incompatibility with materials used in the compressed air system. If petroleum base lubricants are used, synthetics may dissolve the deposits and form a viscous tar, possibly causing damage to the intercoolers, aftercoolers, and valves. Complete cleaning should be done before changing from petroleum base to synthetic lubricant. In addition, be certain the synthetics are compatible with all materials used in the compressor and in all components downstream, especially polycarbonates used in filter bowls. Ask the compressor manufacturer to provide recommendations on any modifications that may be necessary *before* changing to a synthetic lubricant.

ROTARY COMPRESSORS

The early rotary compressors were water cooled, injection lubricated, sliding vane machines. (See Figure 4-2.) They could be used for single-stage low pressure service to 50 psig or in tandem with an intercooler and second stage to pressures of 125 psig. Compact design, low vibration, inexpensive initial cost, and simple design were among the advantages offered versus large reciprocating compressors. The chief disadvantage cited by most plant engineers was the total oil consumption. An external lubricator driven by a V-belt from either the compressor or motor shaft force feeds oil into the air stream and would have to be separated after it left the compressor. In this design, inspection and possible replacement of the vanes was recommended at yearly intervals.

Later designs used the sliding vane compressor with an integral oil reclamation system as part of the design. In this system the oil is used to lubricate, absorb the heat of compression as air is being compressed, and provide a seal between rotors, vanes, cylinder and end plates. Today's rotary machines are primarily of the twin helical screw oil injected design consisting of two rotors meshing in a dual bore cast cylinder (Figure 4-2). The primary configuration is a four-lobe helical male and six-groove helical female screw. The main rotor (usually the male) is driven through a shaft extension by the electric motor while the secondary rotor (usually the female) is driven by the main rotor. There is no metal to metal contact because of the oil film developed by oil injection. The heat of compression is absorbed by the oil, which is water cooled with a shell and tube heat exchanger or air cooled with a fan and radiator assembly. This has been referred to as a flood lubricated compressor and is available as a pre-engineered package (see Figures 4-12, 4-13) with all components mounted and wired to simplify installation. The oil is recirculated and mixed with the air in the compression chamber. The oil and air are separated after compression in an efficient combination receiver and air/oil separator.

> **Note: The twin screw compressor is also available in oil-free design where the rotors are driven through timing gears. Multi-stage units are used for pressures above 50 psig. The advantage of this design is the lack of oil in the compression chamber and in the compressed air. The discussion on rotary compressors is limited primarily to the oil-flooded design.**

Air- and water-cooled units perform equally well and have the same efficiencies. The selection of air- or water-cooled is completely a function of ambient conditions, location, convenience of installation and availability, and economics of water supply versus air cooling. In both air- and water-cooled units, a temperature regulating valve is used to maintain optimum oil temperatures. It is important that

Courtesy of Gardner-Denver

Figure 4-12.

Tank-mounted, air-cooled rotary screw compressor available from 7½ to 40HP.

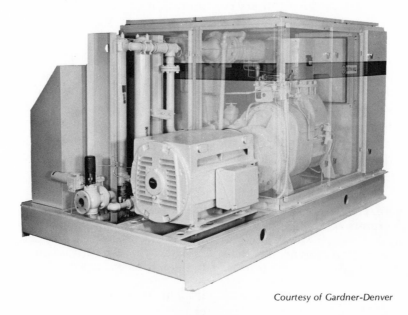

Courtesy of Gardner-Denver

Figure 4-13.

Packaged water-cooled rotary screw compressor available in air-cooled or water-cooled through 500HP.

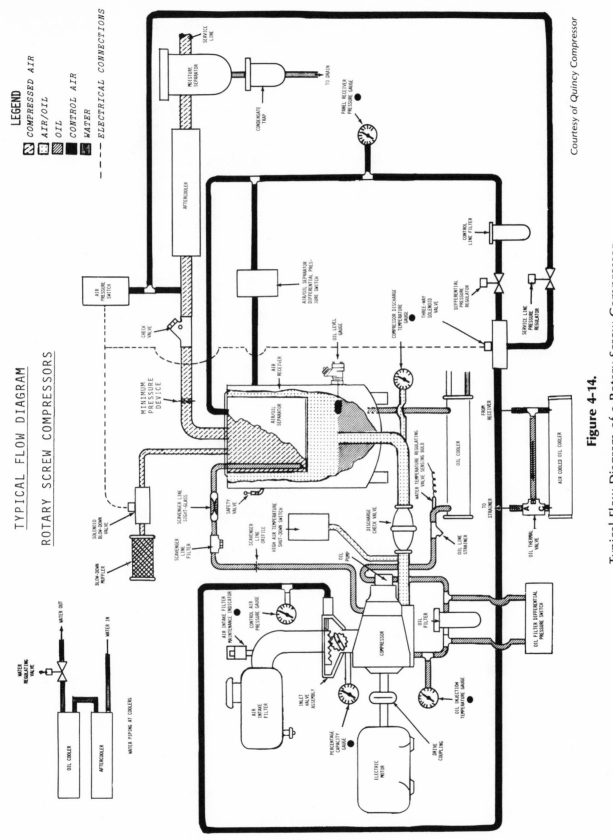

Figure 4-14.

Typical Flow Diagram of a Rotary Screw Compressor.

123

oil temperatures are maintained above the pressure dewpoint, which is approximately 125°F in most installations. The temperature valves are generally set to provide oil at 140°F, which will provide air discharge at approximately 180°F. This is low compared to reciprocating compressors at 100 psig, and thus, carbon and varnish deposits are virtually eliminated. It is important to maintain oil temperatures above the pressure dewpoint to avoid condensation of the air in the oil reservoir. In very humid areas this temperature may be as high as 145°F. However, it is also important to ensure that oils are not subjected to high temperatures, and a high air temperature switch is used to shut the compressor down if discharge air temperatures reach approximately 225°F.

Rotary compressors are designed to run at 100 percent capacity. The method of controlling the capacity is to modulate or to gradually choke the intake as pressure rises, thus reducing the intake air. With this arrangement, the volume is reduced but the pressure ratio across the machine increases, resulting in a relatively flat horsepower curve over the capacity range of the compressor. Most plants should operate a rotary flood lubricated compressor at full load and use a reciprocating unit to provide for peak loads. Some manufacturers of air compressors have introduced a power-saving device built into the rotary screw casing which effectively shortens the rotor lengths at part load. This will reduce the horsepower requirements proportionately to capacity and more closely approximate the reciprocating compressor power characteristics, while retaining the features of the rotary.

Primary Components of the Rotary Compressor (see Figure 4-14.)

A. Intake Filter

This very essential item is especially important in the rotary machine where air and oil are mixed and the oil is then removed in the separator. If the air entering is contaminated, the oil and air mixture is dirty and the separator will require frequent replacement.

B. Oil Separation

The separator is basically a large submicronic filter generally located in the air/oil receiver. Much of the oil is separated in the receiver and the air/oil mixture is passed through the separator to remove most of the remaining oil, and returned to the oil system through a small scavenger line while the air goes to the discharge line. Oil carryover varies from 2.25 ounces per 100,000 cubic feet of free air (and according to some manufacturers) down to perhaps ¼ of an ounce per 100,000 cubic feet.

C. Oil Filtration

The oil filter is an essential component that ensures foreign materials do not enter the compressor.

The above three filters are the primary maintenance items in a flood-lubricated compressor. Proper maintenance of the intake filter and oil filter will extend separator life and will reduce compressor maintenance.

D. Safety and Control Components

1. Thermostatic valve—regulates water flow in water-cooled machines or bypasses oil around the radiator in air-cooled units to maintain 140°F oil temperature.

2. High air temperature switch will shut down the machine at approximately 225°F if air temperature rises.

3. Pressure relief valve is mounted in the air/oil receiver and acts as a safety valve against excessive pressure.

4. A check valve is used after the air/oil receiver to prevent air from coming back to the compressor when the compressor is unloaded or shut down.

5. An automatic solenoid valve will "blow down" or relieve the pressure in the air/oil receiver on shutdown. Some manufacturers also operate this valve each time the compressor fully unloads to reduce the unloaded horsepower.

6. Inlet valve—As explained, this valve is operated by a pilot valve (also called a pilot differential, subtractive pilot, modulating valve, etc.), which gradually opens or closes the inlet valve as pressure rises or falls. Some manufacturers allow the compressor to modulate to zero. Others will modulate to 40 percent of full load capacity and, at that point, will fully unload by closing off the intake completely and simultaneously blow down the receiver. This is done to reduce the no-load horsepower.

7. Instrumentation—There are many devices, gauges, lights, etc., used to indicate performance of a rotary screw compressor. These include, but are not limited to, air and oil temperature gauges, pressure gauges, filter and/or separator warning lights, and air filter indicators. Each manufacturer has different standards and options. Familiarity with the instrumentation will enable the operator to avoid major problems. Record readings periodically and investigate any deviations.

The selection of a proper lubricant must be the responsibility of the manufacturer. Because of the wide variance in requirements and opinions among manufacturers, no recommendations can be given that will cover all rotary compressors. Since manufacturers change their specifications and new oils are being developed, call the manufacturer annually to get the most recent recommendations.

Basic maintenance and trouble-shooting are covered in the charts that follow (Figures 4-15 and 4-16).

ROTARY COMPRESSOR ROUTINE INSPECTION AND SERVICE

Air Filter

Frequency of servicing depends on the environmental conditions under which the compressor is operating.

Service the air filter element under any of the following conditions, whichever comes first:

A. As indicated by the air filter service indicator (located at the instrument panel)

B. Every six months

C. Every 1,000 hours of operation

In some machines filter elements can be cleaned by washing in a warm water solution of household detergent. Discard the element after a maximum of six washings. Keeping a spare filter element on hand is recommended to keep machine downtime to a minimum. Safety element (where applicable) should be changed once a year. Discard the used element.

Air/Oil Separator

Change the separator element under any of the following conditions, whichever comes first:

A. As indicated by the separator warning light (located at the instrument panel) or indicator

B. Once a year

C. Every 4,000 to 8,000 hours (refer to manufacturer's instruction book)

D. If excess oil appears in the service line as a result of faulty separator (due to pin hole, loosened plastisol, etc.)

Oil Filter

Change the oil filter element(s) under any of the following conditions, whichever comes first:

Figure 4-15.

Rotary Compressor Routine Inspection and Service.

A. As indicated by oil filter warning light (located at the instrument panel) or indicator

B. Every six months

C. Every 1,000-2,000 hours or as directed by instruction book

D. Every oil change

Oil Change

Change oil after the first 1,000 hours of operation to make sure the initial contaminants in the system are removed. After that, the oil should be changed every 2,000 hours or in accordance with instruction book; check manufacturer's recommendations if synthetic oil is being used.

Scavenger Line Filter and Orifice (where applicable)

Clean scavenger line filter element and orifice:

A. when no oil flow is visible through the sight glass

B. every 2,000 hours

Control Air Line Filter (Manual)

Condensate collected in the bowl should be drained periodically. Excessive condensate built up in the bowl could adversely affect the performance of the capacity control system.

Oil Cooler

If air-cooled, keep cooling fins clean. If water-cooled, periodic inspection and chemical cleaning may be required.

NOTE: These maintenance and inspection routines are guides. In all cases the manufacturer's recommendations are to be given preference over routines listed above.

Figure 4-15 *(continued)*.

TROUBLE-SHOOTING—ROTARY COMPRESSORS

A. *Failure to Start*
   1. Power failure or power not supplied to starter
   2. Improperly wired
   3. High temperature switch shutdown
   4. Overloads tripped
   5. Control circuit malfunction

B. *Unscheduled Shutdown*
   1. High air temperature caused by:
      a. low oil level
      b. clogged oil cooler or filter
      c. thermostatic valve malfunction
      d. poor ventilation (air cooled only)
   2. Overload Relay Tripped
      a. discharge pressure too high
      b. low voltage
      c. incorrect thermal relay
      d. poor contacts
      e. faulty pressure switch
      f. separator clogged

C. *Low Capacity*
   1. Clogged air filter
   2. Inlet valve not fully open
   3. Pilot valve adjustment required
   4. Too much air demand

D. *High Oil Consumption*
   1. Oil level overfilled
   2. Faulty air/oil separator
   3. Scavenger tube clogged
   4. Incorrect oil
   5. Oil leaks at seals, fittings or gaskets

Figure 4-16.

Trouble-Shooting—Rotary Compressors.

ACCESSORIES

While the trend is towards packaged units with accessories mounted and piped as part of the package, selection of the proper accessories will reduce maintenance problems. In addition, while manufacturers do standardize on the accessories that are suitable for most applications, some problems have been eliminated by changing accessories.

For example, one plant had many lubricated reciprocating compressors and was constantly replacing aftercoolers with the original type provided. After examination it was determined that the cooler had small tubings with an internal fin that would quickly plug with carbon deposits. Changing to an aftercooler made by the same manufacturer, but with larger unrestricted tubes, solved the problem. Another plant had continuous problems with the separators in the oil-flooded screw compressors. Upon investigation it was noted that a very fine cosmetic dust was getting through the intake filter and collecting in the separator, causing frequent costly replacement. Replacing the intake filter with one that had a finer filtration and a larger element solved the problem.

Aftercoolers

Reducing moisture in compressed air is a problem most plant engineers face at one time during their careers. The basic accessory for moisture removal is the aftercooler. By reducing the temperature of the air, the water vapor is condensed and can be separated and drained from the system. In addition, the aftercooler provides a degree of safety in preventing explosions. If carbon accumulated in the compressor discharge of a reciprocating compressor and it became sufficiently incandescent, the fire would probably be quenched as it entered the aftercooler, precluding the possibility of a major explosion if the fire reached the air receiver.

Aftercoolers are rated in cfm at a given pressure and the approach temperature. In the case of a water-cooled shell and tube aftercooler, the approach temperature is the difference between the incoming water temperature and the outgoing air temperature. Thus, a cooler with water entering the shell at 85°F and air leaving the tubes at 100°F would have a 15°F approach. Check the approach temperature with the manufacturer and the amount of water required to cool. This may vary from one gallon to four gallons per 100 cubic feet. Also, check tubing size and ease of cleaning the tubes when ordering aftercoolers. Install indoors to prevent freezing and leave room to pull the tube bundles for cleaning and maintenance. Clean, soft water is essential; and the cooler, the better. Warm water results in warmer air with more vapor left in the air, and also results in accelerated scaling in the jacket. Check the tubes and jackets yearly and clean as required in accordance with the manufacturer's instructions.

Air-cooled aftercoolers are generally radiator-type units that cool the air to an approach temperature from 10°F to 25°F of the ambient temperature. In a high ambient temperature an air-cooled aftercooler may be unsuitable. Some air-cooled aftercoolers are cooled by the flywheel of an air-cooled compressor while others have a separate fan. Be sure to check the approach temperature of the aftercooler and do install it in a cool, clean location. Maintenance consists of cleaning the fins with a compressed air blowgun. If the fins are oily, a nonflammable solvent may also be necessary.

Separators should be purchased with aftercoolers; and as the name implies, they separate the water from the air. An automatic drain trap is installed at the bottom of the separator to drain accumulated moisture.

Hints

1. On installations where large aftercoolers are not purchased as part of a package, a three-valve bypass will enable maintenance to be done on the aftercooler while the compressor continues to operate. Be certain to relieve all pressure in the aftercooler and shut off and drain all water before attempting repairs or maintenance.

2. A safety valve before the aftercooler is useful if carbon buildup could be a problem. If the cooler begins to plug, the safety valve will pop.

3. Place a manual shutoff valve before any drain trap so that it can be repaired without shutting down the system.

4. The outlet from an automatic drain trap should be visible to the operating engineer. Visibility is necessary to determine if the trap is working. Either pipe to an open funnel or have gauge glasses installed on the separator to see if water accumulates. There should not be water in the separator.

5. Check automatic drain traps manually to be certain they are working.

Air Receivers

1. Be certain the air receiver is of ASME construction and that the stamped working pressure is well above the operating pressure. An ASME safety valve should be installed and set at least 5 psi below the working pressure of the receiver.

2. Receivers should be drained frequently to be clear of condensate, oil, scale, etc.

3. As a rough rule of thumb, receivers are sized so that the capacity of the air compressor in cfm divided by three is the size in cubic feet of the receiver for automatic start-stop control. The cfm divided by seven is generally used for constant speed control. For rotary compressors operating on modulating controls, the receiver can be very small and in some applications is not required.

The proper selection of one or more tanks is based not only on the capacity of the compressor itself, but also on the shop load cycle. The following formula can be used to determine the correct size:

$$V = \frac{T(C-Cap)(P_b)}{(P_1-P_2)}$$

T   = Time interval in minutes, during which a receiver can supply air without excessive drop in pressure

V   = Volume of tank in cubic feet

C   = Air requirement in cubic feet of free air per minute

Cap = Compressor capacity

P_b = Absolute atmospheric pressure, psia

P_1 = Initial tank presure, psig (compressor discharge pressure)

P_2 = Minimum tank pressure, psig (pressure required to operate plant)

If Cap is greater than C, the resulting negative answer indicates that the air compressor will supply the required load. If the compressor is unloaded or shut down, Cap becomes zero and the receiver must supply the load for "T" minutes.

Intake Filters

Air entering any compressor must be filtered. Foreign matter drawn into the machine will cause rapid wear and will also contaminate the lubricant.

For nonlubricated compressors or where extremely fine filtration is necessary, a good dry-type filter with either a felt cloth or special paper media is recommended. The felt type can be cleaned, whereas the paper type is replaced when dirty.

For lubricated compressors, a viscous impingement-type or oil-bath filter may be acceptable. The viscous impingement type has woven wires packed into a cylinder or cell that is coated with oil to hold dirt. This should not be used in a dusty area. The oil-bath filter is preferred over the viscous impingement and removes dirt by washing it down a fine strainer into a sump from which it is manually drained. The oil must be changed and the filter cleaned as conditions require. Clogged filters are responsible for many problems besides wear on the compressor. A dirty filter reduces the volume of air compressed, increases the pressure differential across the machine, causes overheating, and may even collapse and be drawn into the compressor.

Some filters are manufactured with a silencing chamber to reduce noise level. There are also separate intake silencers available which, when installed close to the compressor intake, will often reduce noise levels. When air is drawn from outside the compressor location, increase the pipe one inch for every 10 feet of length and use a full-size filter with a rain hood if the filter is outdoors. Check with compressor manufacturer on lengths to avoid resonance in intake pipe.

There have been cases where intake filters were located near soot blowers and steam exhausts, and compressor life was substantially reduced. In other cases the intake was near paint fumes, which was not only detrimental to compressor life, but posed a definite safety hazard as well. Be certain that wherever the intake is located, it will draw clean, cool, dry air.

Reducing Compressor Maintenance by Proper Installations and Start-Up Procedures

1. Read the instruction book thoroughly.
2. Is the intake piped to a clean location?
3. Be certain there is adequate cooling.

   a. Air cooled—Adequate ventilation and cooling
      —Clean, dry air

   Remember, an air compressor will dissipate 2545 BTU/hour of operation. On an air-cooled unit, this will add to the ambient temperature already in the area unless adequate ventilation is available.

   b. Water-cooled—Is there adequate water at the proper pressure?

   Can the water be drained to a sewer or must a cooling tower be considered?

   Must the water be treated?

4. Is there adequate power *at* the compressor location?

   Does the voltage on the motor nameplate agree with the supply?

   Is the correct magnetic starter installed to meet inrush current limitations?

5. Is a foundation necessary? If yes, what soil bearing pressure can be tolerated?

6. At start-up:

   a. Proper lubricant and at the correct level?

   b. Machine level?

   c. Coupling aligned or belts adjusted properly?

7. a. Turn machine over by hand.

   b. Press button to jog compressor and check rotation.

   c. Listen for any unusual noises.

   d. Run the compressor—unloaded for period of time, if recommended.

8. Measure and record:

   a. Full load current and voltage.

   b. Time to pump air receiver to cut out pressure.

For evaluations and considerations on what type of air compressor should be purchased, see Figure 4-17.

The basic types of compressors and some of the parameters to evaluate are as follows:

Type	Initial Purchase Price	Installa-tion	Repairs	Mainten-ance	Power Cost (Full Load)	Power Cost (Part Load)
Reciprocating						
1. Air-cooled	1	1	4	3	4	2
2. Water-cooled single-acting	2	2	3	3	2–3	2
3. Water-cooled double-acting:						
Single-stage	3	4	2–3	3	2–3	1
Two-stage	4	3*	2–3	3	1	1
Rotary Screw	1–2	1	1	1	2–3	4**

KEY: 1. Very Good   2. Good   3. Fair   4. Poor

It is well to consider the packaged concept where the manufacturer supplies all components prepiped and prewired to facilitate installation. Most rotary screw compressors and smaller air-cooled reciprocating compressors are sold packaged.

*A two-stage unit generally requires smaller foundations than single-stage. However, rigging costs may be higher.

**Some rotary screw compressors with a power-saver feature can have good part load characteristics and can be rated 2.

Other factors to consider that can influence purchase include:

- Floor space
- Noise level
- Ventilation requirements
- Quality of air required
- Subsoil conditions
- Availability of qualified maintenance

Figure 4-17.
Types of Compressors and Parameters of Their Evaluation.

Air Dryers

There are three basic types of air dryers—deliquescent, regenerative desiccant, and refrigerated. (See Figure 4-18.)

Deliquescent Dryers

Deliquescent dryers consist of a receiver filled with a chemical desiccant that reacts with moisture in the air and absorbs it by dissolving the surface desiccant. The water-salt solution drains to the bottom of the receiver where it must be drained. This type of dryer requires the desiccant to be replenished. The inlet temperature must be limited to 100°F or less to avoid increased consumption. The effectiveness of this dryer is questionable under certain conditions and while it does have a low initial cost, it is no longer considered the prime method of eliminating moisture in a compressed air line. A dewpoint suppression of 20°F is obtainable.

Regenerative Desiccant Dryers

Regenerative desiccant dryers consist of two towers, each containing a charge of a solid desiccant such as silica gel, activated alumina, or molecular sieve. The air flows alternately through each tower so that air is being dried or regenerated either by heaters or dry air is being purged back through the desiccant.

The regenerative dryers can produce dewpoints of −40°F to −100°F depending on the desiccant material selected. With purged air being used for regeneration, the cycle is very short (5 to 10 minutes) and the purged air can be 10 to 15 percent of the total supplied, so that this reduces the total air available. With a heater type dryer, the cycle is four to eight hours and electric heaters (or steam or hot air) are used. Consider total energy required for regeneration when evaluating this type dryer.

Regenerative dryers should be sized quite closely to the requirements to avoid wasteful purging or excessive heater operation and replacement. The desiccant will probably have to be replaced yearly in most cases.

The deliquescent and regenerative dryers have desiccant that will not remove the moisture if it is coated with oil. Therefore, in a lubricated system, put an oil removal filter before the dryer. Since desiccant dusting can occur, a fine filter after the dryer will prevent desiccant from entering the system.

Refrigerated Dryers

The refrigerated dryer has almost become the standard to eliminate moisture in manufacturing plants where the dewpoint required is above 35°F. Most manufacturers use a noncycling type of dryer that has an air-to-air and refrigerant-to-air heat exchanger packaged to include instrumentation, automatic drain trap and

Typical Refrigerated Dryer

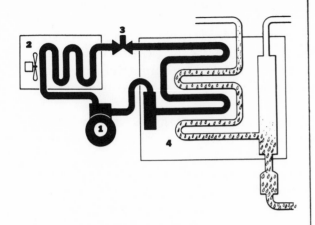

MAJOR COMPONENTS OF THE REFRIGERATED AIR DRYER

1. **Refrigeration Compressor:** A hermetically sealed motor driven compressor operates continuously. It generates a high pressure refrigerant gas.

2. **Hot Gas Condenser:** The high pressure refrigerant gas enters an air cooled condenser where it is partially cooled by a continuously running fan.

3. **Automatic Expansion Valve:** The high pressure liquid enters an automatic expansion valve where it thermo-dynamically changes to a subcooled low pressure liquid.

4. **Heat Exchanger:** A system of coils where dry air is produced.

Typical Regenerative Dryer
[Heatless type - also available with heaters for desiccant regeneration]

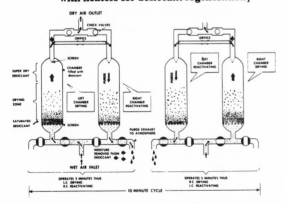

ACTUAL OPERATION — As shown in the above flow diagrams, the heatless Dryer cycles between two desiccant chambers, one serving as a drying medium, while the other is undergoing a reactivation process. The diagram at left shows the wet gas entering at the bottom of the left-hand chamber, passing upward through the desiccant where it is dried to an extremely low dewpoint. The dry air passes through the check valve to the dry air outlet. Simultaneously, a small percentage of the dry gas is expanded through the orifice between the chambers, and flows down through the right-hand chamber, reactivating the desiccant, and passing out through the purge exhaust. At the end of the cycle, the chambers are automatically reversed, the right-hand chamber serving as the drying medium, while the left-hand chamber is being reactivated, as shown in the diagram at right.

Typical Desiccant Dryer

ATMOSPHERIC DRYNESS

DEW POINT °F	PARTIAL PRESS MM HG	GRAINS MOISTURE # AIR	# MOISTURE # AIR	GRAINS MOISTURE CU. FT. AIR	PPM	% R H a RM TEMP.	VOL %
110		400	.0600	25	60.000		9
			.0500		50.000		8
							7
100		300	.0400	20	40.000		6
			.0300		30.000		5
90		200		15			4
80	24.0	150	.0200	10	20.000	100.0	3
	22.0			9		91.7	
	20.0		.0150	8	15.000	83.3	
70	17.5	100		7		72.9	2
	15.0	90		6		62.5	
60	12.5	80	.0100	5	10.000	52.1	1.5
		70	.0090		9.000		
		60	.0080		8.000	41.7	
50	10.0		.0070	4	7.000	37.5	1
	9.0	50	.0060		6.000	33.3	9
	8.0					29.2	8
	7.0	40	.0050	3	5.000	25.0	7
40	6.0		.0040		4.000	20.8	6
	5.0	30		2			5
30	4.0	20	.0030	1.5	3.000	16.7	4
	3.0		.0020		2.000	12.5	3
20				1			2
	2.0	10	.0010	.8	1.000	8.3	
10		8	.0008	.6	800		1
	1.0	6	.0006		600	4.2	.08
0		4		.4			.06
	.5		.0004		400	2.1	.04
−10	.32	2		.2		1.3	.02
−20			.0002		200		
	.178	1		.1		.74	
−30		.8	.0001	.08	100		
	.095	6	.00008	.06	80		.01
−40		4	.00006	.04	60		.008
	.050		.00004		40		.006
−50		2		.02			.004
	.026		.00002		20		
−60		1		.01			.002
	.013	.8	.00001	.008	10		
−70		.6	.000008	.006	8		.001
	.0062	.4	.000006	.004	6		.0008
−80			.000004		4		.0006
	.0028	.2		.002			.0004
−90			.000002		2		
	.0012	.1		.001			.0002
−100							

proper controls, to match the compressed air load to be dried to available refrigeration capacity. Larger chiller-type dryers use a closed circuit water chiller which cools a mass of water which is circulated through an aftercooler. The compressed air passes through this aftercooler and water vapor is condensed as it would be in a standard aftercooler.

Some dryers require pre-filters because the heat exchangers have very small passages that may easily plug with contaminants such as rust, scale, or carbon from a compressed air system.

The primary consideration in selecting a refrigerated dryer to remove moisture is to understand the design criteria so that an evaluation can be made that will prevent problems from occurring when the dryer is placed in service. The parameters to consider include:

- inlet air temperature
- ambient air temperature
- cfm to be cooled at pressure
- pressure drop
- pressure dewpoint

There have been cases where a dryer purchased with particular design criteria did not operate on a hot summer day when it was needed most.

For example, most dryers are rated at 100°F inlet air temperature operating in a 100°F ambient. At temperatures other than 100°F, the capacity is increased or decreased as follows:

Inlet Temperature, °F	Capacity Multiplier
90	1.23
95	1.13
100	1.0
105	.9
110	.83
120	.69

Ambient air temperatures have a similar effect but not as marked.

Ambient Air Temperature, °F	Capacity Multiplier
85	1.09
90	1.05
100	1.0
110	.91

A dryer rated for 100°F ambient air and 100°F inlet air would have far greater capacity than one rated at lower temperatures. Conversely, and more importantly, the purchase of a dryer rated at 1,000 cfm with air entering at 90°F in an ambient of

85°F would be capable of drying only 746 cfm at 100°F inlet air and 100°F ambient temperature.

Pressure drop is normally 5 psi maximum, but remember that this will vary as the square of the cfm through the dryer. Thus, increasing the cfm by 10 percent will increase the pressure drop by 21 percent. Also, changes in operating pressure affect the pressure drop inversely as the ratio of absolute pressures. If the pressure drop is 5 psi with 1,000 cfm flowing at 100 psig (114.7 psig), the pressure drop at 50 psig (64.7 psig) would be $\frac{114.7}{64.7} \times 5$ or 8.9 psi.

The pressure dewpoint of 35°F is one of the standards most commonly accepted, but 50°F is also used. This becomes a plant engineer's choice as to which is best suited for his plant.

Basic maintenance of dryers consists of monitoring refrigeration gauges to insure operation of the refrigeration compressor, checking the automatic drain trap and repairing when necessary, and keeping the condenser clean. Cleaning the compressed air side of the heat exchangers can be done in many dryers by back flushing air in reverse direction through the dryer, or in some cases, by pumping an acceptable solvent through the dryer. Check the manufacturer's instructions for detailed instructions. If the heat exchangers become dirty, the pressure drop across the dryer will increase. Always install a three-valve bypass around the dryer to facilitate maintenance when required and to ensure uninterrupted air supply.

Oil removal filters used in conjunction with air dryers have helped produce "instrument quality" air where lubricated compressors have been installed. Where no oil can be tolerated, such as use of air in a food product, it is generally best to use non-lubricated compressors. It is recommended that pressure gauges be installed before and after dryers and filters to monitor pressure drop that when excessive, will indicate a maintenance problem.

Energy-Saving Hints

1. Fix all leaks.
2. Operate the air compressor at the lowest possible pressure that will still allow all equipment to be run. Some compressors will save 1 percent in power for every 2 psi pressure reduction so that a 10 psi drop could result in a 5 percent power savings. A second benefit is that compressors generally last longer at lower pressures. (Rotary flood lubricated compressors have a pressure below which they should not operate unless specifically designed for low pressure.)
3. Pipe intake to a clean, cool location. Reciprocating compressor capacity can be increased 1 percent for every 5°F drop in temperature with no increase in energy consumption. (Rotary compressor capacity is basically unaffected by changes in inlet temperature.)
4. Heat recovery from air compressors has become an excellent method of conserving energy.

a. In water cooled compressors, water discharged from an air compressor at 110-125°F can be used to feed processes that had been heating water from municipal water supplies. One food processor required huge quantities of hot water for cleaning and was able to use the hot discharged water through a heat exchanger to heat pure cleaning water.

b. In air-cooled compressors, rotary screw compressors have become especially adaptable to use the dissipated heat from the oil cooler and aftercooler to heat an area for many months of the year. During warmer months, the hot air is vented to the outside through duct work. Don't forget to consider the cost of water in computing water-cooled costs and the fan horsepower in considering air-cooled energy costs.

The Pipeline System

Figure 4-19 provides friction losses through pipelines. This should be used to size a distribution system. Some hints that have helped plant engineers design a good system include:

1. Size main lines for maximum flow with a maximum 2 percent pressure drop.

2. Size branch lines for average flow with 2-3 percent pressure drop.

3. Pitch all lines to a low point at one inch per 20 feet. Place drains at the low points.

4. Take all compressed air outlets from the top of the pipe so that any condensation, scale or contaminants remain in the pipeline and can be drained off at the low points.

5. Where threaded pipe is used, use tees with the branch plugged rather than couplings to join two pieces of pipe. This allows future expansion and branches if necessary without the necessity for cutting into a pipeline.

6. Consider expansion when designing a system so the pipelines are large enough to supply additional equipment.

7. If a certain piece of equipment uses a large volume of air suddenly, intermittently but repetitively, a surge receiver at the point of use will prevent pressure fluctuations in the line.

8. Remember that pressure drop in a pipe varies as the square of the volume flow. For example, 1,000 cfm of free air in a four-inch pipe has a pressure drop of 2.21 psi in 1,000 feet of pipe at 100 psig. Two thousand cfm will have a pressure drop of 8.80 psi in the same pipe. Also, if the pressure drop at 50 psig (64.7 psia) is required in the same four-inch pipe with 1,000 cfm of free air, it would be inversely proportional to the absolute pressures. Therefore, the pressure drop would be $\dfrac{114.7}{64.7} \times 2.21 = 3.92$ psi.

AIR TRANSMISSION – Friction Losses Thru Pipe

Delivery in Cu. Ft. of Compressed Air per Min.	Equiv. Delivery in Cu. Ft. of Free Air per Min.	1/2	3/4	1	1-1/4	1-1/2	2	2-1/2	3	3-1/2	4	4-1/2	5	6	8	10	12
					SIZE OF PIPE, INCHES — LOSS OF AIR PRESSURE DUE TO FRICTION (in psi in 1000 ft.* of pipe)												
					At 100 Pounds Gauge												
5.12	40		16.0	4.45	1.03	.46											
8.96	70		49.3	13.7	3.16	1.40	.37										
12.81	100			27.9	6.47	2.86	.77	.30									
15.82	125			48.6	10.2	4.49	1.19	.46									
19.23	150			62.8	14.6	6.43	1.72	.66	.21								
22.40	175				19.8	8.72	2.36	.91	.28								
25.62	200				25.9	11.4	3.06	1.19	.37	.17							
31.64	250				40.4	17.9	4.78	1.85	.58	.27							
38.44	300				58.2	25.8	6.85	2.67	.84	.39	.20						
44.80	350					35.1	9.36	3.64	1.14	.53	.27						
51.24	400					45.8	12.1	4.75	1.50	.69	.35	.19					
57.65	450					58.0	15.4	5.98	1.89	.88	.46	.25					
63.28	500					71.6	19.2	7.42	2.34	1.09	.55	.30					
76.88	.600						27.6	10.7	3.36	1.56	.79	.44					
89.60	700						37.7	14.5	4.55	2.13	1.09	.59					
102.5	800						49.0	19.0	5.89	2.77	1.42	.78					
115.3	900						62.3	24.1	7.60	3.51	1.80	.99					
126.5	1000						76.9	29.8	9.30	4.35	2.21	1.22					
192.2	1500							67.0	21.0	9.80	4.90	2.73	1.51	.57	.24		
256.2	2000								37.4	17.3	8.80	4.90	2.72	.99	.24		
316.4	2500								58.4	27.2	13.8	8.30	4.20	1.57	.37		
384.6	3000								84.1	39.1	20.0	10.9	6.00	2.26	.53		
447.8	3500									58.2	27.2	14.7	8.20	3.04	.70	.22	
512.4	4000									69.4	35.5	19.4	10.7	4.01	.94	.28	
576.5	4500										45.0	24.5	13.5	5.10	1.19	.36	
632.4	5000										55.6	30.2	16.8	6.30	1.47	.44	.17

*For longer or shorter pipes the friction loss is proportional to the length, i.e. for 500 feet 1/2 of the above; for 4,000 feet four times the above, etc.

LOSS OF PRESSURE THROUGH SCREW PIPE FITTINGS
(Given in equivalent lengths (feet) of straight pipe)

Nominal pipe size, (inches)	Actual inside diameter, (inches)	Gate valve	Long radius ell or on run of standard tee	Standard ell or on run of tee reduced in size 50 per cent	Angle valve	Close return bend	Tee through side outlet	Globe valve
1/2	0.622	0.36	0.62	1.55	8.65	3.47	3.10	17.3
3/4	0.824	0.48	0.82	2.06	11.4	4.60	4.12	22.9
1	1.049	0.61	1.05	2.62	14.6	5.82	5.24	29.1
1-1/4	1.380	0.81	1.38	3.45	19.1	7.66	6.90	38.3
1-1/2	1.610	0.94	1.61	4.02	22.4	8.95	8.04	44.7
2	2.067	1.21	2.07	5.17	28.7	11.5	10.3	57.4
2-1/2	2.469	1.44	2.47	6.16	34.3	13.7	12.3	68.5
3	3.068	1.79	3.07	6.16	42.6	17.1	15.3	85.2
4	4.026	2.35	4.03	7.67	56.0	22.4	20.2	112
5	5.047	2.94	5.05	10.1	70.0	28.0	25.2	140
6	6.065	3.54	6.07	15.2	84.1	33.8	30.4	168
8	7.981	4.65	7.98	20.0	111	44.6	40.0	222
10	10.020	5.85	10.00	25.0	139	55.7	50.0	278
12	11.940	6.96	11.0	29.8	166	66.3	59.6	332

Figure 4-19.

A proper compressed air piping system is shown in Figure 4-20.

If good selection and installation procedures are followed, many problems with air compressors, accessories and the compressed air system can be eliminated and maintenance substantially reduced. Where repetitive problems occur, it is very possible that the equipment selected is not correct for the application.

Operating logs and records are essential in establishing a maintenance program that will pinpoint problem areas and reduce operating costs. This can be accomplished with a commitment to evaluate problem areas and work towards eliminating them.

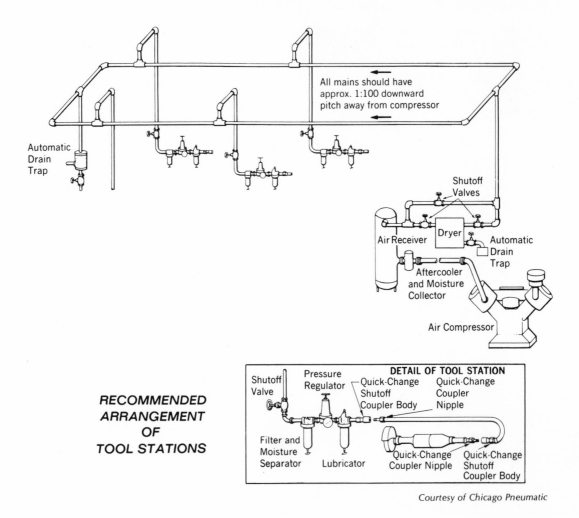

Courtesy of Chicago Pneumatic

Figure 4-20.

5

Effective Motor and Automatic Control Maintenance

Evans J. Lizardos, P.E.

SCOPE OF CHAPTER

This chapter provides you with an understanding of available methods and means to secure better maintenance and operation of motor and automatic controls. You are guided through a two phase step-by-step explanation of basic control examples to develop insight into these systems. These two phases are:

- Identifying typical automatic control components, including motor controllers and types of control systems available.

- Explaining the fundamentals of ladder diagrams to show how this elementary control logic technique can be used to describe and to understand a variety of control systems from the simple to the complex and sophisticated (computer) systems.

You are introduced to, and progressively become familiar with, control circuit elements and the logic of their use. As progress is made through the chapter, confidence is developed gradually in understanding and evaluating control systems. From this confidence, capabilities can be developed to understand advanced degrees of sophistication and capability in other control systems.

An Essential Aid

The elementary logic contained in the ladder diagrams in this chapter is the key to identifying the function and sequence of operations for motor controllers and other types of automatic control components. For those in maintenance, it is also the key to successful trouble-shooting and repairs. Once understood, servicing of control components and overall system circuits is significantly simplified. Therefore, this chapter is dedicated to developing a basic understanding of the logic underlying simple and complex control systems.

First, however, a few sound maintenance program fundamentals for automatic control systems are included in the following elements:

1. Planned Maintenance

   a. On a regular periodic basis, inspect all system control components including all system connections, wire insulation, contact surfaces, basic parts, and other devices.

   b. Test all system components in accordance with manufacturer's instruction and at recommended frequency.

      ■ At least once a year, clean terminals, control boxes, and other parts with compressed air or by vacuum cleaning.

      ■ Simultaneously, check for loose connections.

   c. Adjust or fine-tune system components to reduce the chance of malfunction in vibration environments. Normally adjustment is unnecessary until the control circuit malfunctions.

   d. Replace components in accordance with manufacturer's recommendations. Promptly overhauling the removed components will help ensure an inventory of readily usable units.

   e. Run through logic of the entire system.

2. Failure Response

   Failure frequently occurs during emergency situations. Consequently, it is best to prepare with an adequate inventory of spare parts, relays, contacts, spare printed circuits, switches, etc. It is also prudent to have readily available the manuals for operation, service, maintenance, and spare parts. In addition, keep a posted list of current telephone numbers of key maintenance personnel and emergency service companies, as well as the telephone numbers of manufacturers and spare parts suppliers.

Understanding Control Systems

There are three basic techniques used to describe control circuits: wiring diagrams, written operational sequences, and logic diagrams.

1. Wiring Diagrams

   ■ Wiring diagrams illustrate, by the use of technical symbols and drawing conventions, the physical arrangements and complete electrical wiring requirements. All wiring connections to and from controls, devices and terminals on equipment are located and identified in these diagrams.

   ■ Wiring diagrams are required for a complete and proper installation of control equipment. These diagrams enable electrical personnel involved with system design, installation, cost estimates, etc., to determine location, type, and number of devices and the extent of wiring to be provided. Essentially, these diagrams are shop or field installation drawings. They supply the mechanical and electrical contractors with information required to locate and install equipment in accordance with manufacturer's requirements or engineer's design. However, these wiring diagrams assist operating and maintenance personnel only partially in understanding how the overall system operates or how the individual control device functions. For a more complete understanding there is need for a description of the control system operation and sequence.

2. Operational Sequences

   - An operational sequence can best be described as a written technical explanation of the established operating sequences and prescribed performance of the control and monitoring system.

   - Usually an operational sequence includes lengthy and extensive descriptions to define properly the intended needs and logic of the system. As these descriptions extend into long and complex explanations, comprehension becomes difficult for some people and impossible for others. As a result, there is a need for a third control explanation technique to convey technical understanding. This is the logic diagram that is often available but seldom used, except by those involved in the original control system design.

3. Logic Diagrams

   - Logic diagrams have been in existence as long as wiring diagrams. Nevertheless, they have received little use because few operating and maintenance personnel are aware of their advantages. Wiring diagrams and operational sequences satisfy most maintenance needs, especially when control circuits are simple. In recent years, though, these circuits have become more complex and increasingly difficult to understand. Control equipment manufacturers have increased the number of operating modes and introduced more safety features in their equipment. As a consequence, manufacturers have recognized the value of logic diagrams, and are now actively promoting their use.

   - The logic diagram is extremely valuable in understanding how a control circuit operates. Its greatest attribute is the ease with which one gains an understanding of the step-by-step logic. It is especially valuable as programmable controllers and computers become more widely used.

   - Logic diagrams are also called elementary, fundamental, or ladder diagrams. In this chapter, they are referred to as ladder diagrams, because each step of the logic builds on the previous steps, like rungs in a ladder.

Ladder Diagrams

Ladder diagrams are essentially the identified steps of a control logic. As mentioned above, these steps are shown graphically as rungs in the ladder diagram.

The predetermined steps of control logic are usually displayed as inverted steps in a ladder. They start from the top with first step (Rung No. 1) and work down on the ladder diagram. The first step usually starts with the power source for the control system. Each subsequent step adds one or more elements to the control circuit in the sequence they would normally occur, or in the sequence that is easy to understand. To conserve space, standard symbols and abbreviations are used on each rung of the ladder; however, they convey a complete control message. A written explanation of the control logic for one complex rung could require several paragraphs of text. The symbols and notations on the rung convey the same information.

Ladder diagrams have several applications, but the three most common uses are identified as follows:

1. Accurate trouble-shooting: The maintenance man who understands the control logic can pinpoint control system problems faster than one who does not understand the function. On a production line, this can often make the difference between minutes and hours of downtime; long delays in production operations can be disastrous. A knowledgeable maintenance technician avoids erroneous assumptions about where the control fault is located. The ladder diagram assists in localizing the fault area and also permits the trouble-shooter to recognize the cause of the malfunction.

2. Add to existing system: Additions to existing control systems require more than adding wire and components. Additions must be totally compatible with system logic and the physical installation constraints. Since there are several ways to wire a complex circuit, the best arrangement can be identified on paper through modifications and extensions of the original logic diagram.

3. Understanding sophisticated solid-state controllers: Without a logic diagram, the complex circuits inherent in solid-state controllers would be nearly impossible to follow and comprehend. Since programmable controllers are rapidly becoming popular, it is incumbent on those in maintenance to be prepared to service an increasing number of these sophisticated systems.

The application of ladder diagrams, for understanding control logic, permits the maintenance person to readily grasp the systems logic and intent, thereby enabling him to understand the sequence of operation. Trouble-shooting a control circuit is considerably simplified when the maintenance person understands how the system functions.

The underlying principles used in logic diagrams are the same for electric, pneumatic, and other control circuits. However for brevity and convenience, this chapter explains logic primarily in electric control circuits. The approach to understanding other automatic control systems is identical. This includes, but is not limited to, the following types of control circuits: electric, electronic (including solid-state), and pneumatic. This commonality is illustrated in the application section of this chapter.

Source of Information

The basic source of ladder diagram information is the equipment manufacturer who supplies a variety of pertinent information including:

1. Sales literature and data sheets
2. Engineering details and specifications
3. Equipment manuals
4. Wiring diagrams (frequently mounted somewhere on the equipment)
5. Written description of operations
6. Manufacturers' ladder diagrams

Other sources of ladder diagrams and related information can be obtained from engineers or others responsible for design and installation of control equipment or systems. Their inputs would include ladder diagrams, shop drawings, contract drawings, system manuals, and other related material.

Typical Control Equipment Hardware and Explanation of Selected Relevant Terms

Primarily for economic reasons, most electric motors over ½ horsepower are powered by three-phase alternating current; and motors under ½ horsepower are usually served by single-phase circuits. Fig. 5-1 summarizes alternating current characteristics most frequently used to control and power equipment in the United States.

SUMMARY OF ELECTRICAL CHARACTERISTICS IN THE U.S.

Function	Voltage	Amperes	Phase	Hertz
Control	24 to 120	5 to 15	Single	60
Power for Small Equipment	120 208 240 277	10 to 40	Single	60
Power for Large Equipment	208 240 480 575 or 600	30 to 400	Three	60

Figure 5-1.

Summary of Electrical Characteristics in the U.S.

Figure 5-1 indicates control circuitry is always at 120 volts or less. Consequently, control equipment and devices are designed to be powered at relatively lower voltages. To facilitate explanation and understanding of control systems, the more common types of control equipment, components and devices are identified in the following pages. Where appropriate, selected photographs are used to identify and aid in the description of the equipment. In addition, terms necessary to understand control circuit logic are also defined in this section. These descriptions and equipment identifications are arranged in a building block sequence. The explanations begin with the simplest control terms and equipment, followed by definitions of the progressively more complex components.

1. Control voltage, control power, control circuit
    - All three power source terms mean the same thing. They denote a 120 volt (or less) single-phase electric power source. Most regulatory agencies are adopting 120 volts as a maximum control voltage. They also require the electric source to be grounded.

2. Line voltage (power)
    - This term identifies the electric power source that is 120 volts or greater. It may be single- or three-phase. Line voltage is usually associated with the voltage supplied by the utility for equipment and systems found throughout the plant or building.

3. Switches, disconnects, and manual starters
    - Switches and disconnects are used to interrupt manually a single- or three-phase electric supply to equipment motors and systems. They can be used in control or power circuits. Manual starters are similar to switches and disconnects, except that starters contain a motor thermal overload control device. This overload device automatically interrupts power to the motor when the current draw exceeds the motor's "must trip" ampere rating.

        a. Figure 5-2 shows a simple manual starter with thermal overloads for motor protection. This three-pole starter is shown with an enclosure. The left side screw terminals connect the starter to the motor circuit.

        b. Figure 5-3 shows a motor pushbutton station housing switch push-buttons that control the motor circuit.

        c. Figure 5-4 shows a three-phase manual motor starter that contains start-stop pushbuttons, a three-pole switch, and a thermal overload device.

4. Magnetic or solid-state type relays, contactors and starters
    - Magnetic-type relays and contactors function in the same way. They consist of a magnetic coil that, when energized, pulls together pairs of contacts to complete a circuit. This permits the flow of line power to equipment.

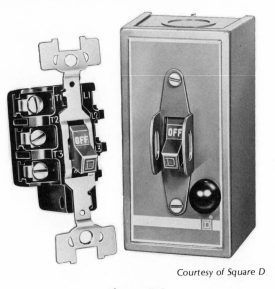

Courtesy of Square D

Figure 5-2.

Manual Motor Starter.

Courtesy of Allen-Bradley

Figure 5-3.

Motor Start-Stop Station.

Courtesy of Furnas Electric Company

Figure 5-4.

Three-Phase Start-Stop Manual Motor Starter.

- Magnetic starters are contactors equipped with thermal motor overload protection devices. These are properly referred to as magnetic motor starters. Figures 5-5 and 5-6 picture representative magnetic motor starters. Figure 5-7 illustrates a cutaway view of a typical thermal overload relay found in a magnetic motor starter.

- Relays and contactors are similar devices. Relays are frequently used in control circuits at 120 volts or less while contactors are identified with line voltage circuit applications. Consequently, the current capacity rating of control relay contacts is usually less than 10 amperes at 120 volts. On the other hand, contactor ratings match the ampere rating of the power-consuming equipment on the line. Figure 5-8 contains typical magnetic relays and contactors shown without enclosures.

- Solid-state relays, contactors, and starters perform the same functions as electro-mechanical relays, contactors, and starters. However, instead of contacts moving mechanically, solid-state circuits are employed to engage and disengage the controlled equipment, component, or device.

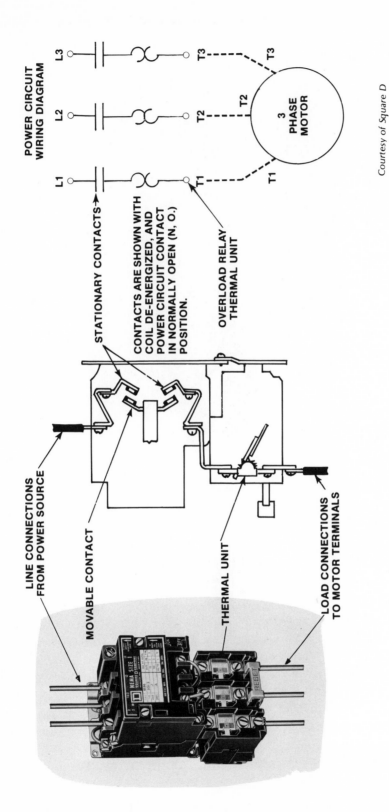

POWER CIRCUIT
WIRING DIAGRAM

L3   L2   L1

T3   T2   T1

T3   T2   T1

3
PHASE
MOTOR

Courtesy of Square D

STATIONARY CONTACTS→

CONTACTS ARE SHOWN WITH
COIL DE-ENERGIZED, AND
POWER CIRCUIT CONTACT
IN NORMALLY OPEN (N, O.)
POSITION.

OVERLOAD RELAY
THERMAL UNIT

LINE CONNECTIONS
FROM POWER SOURCE

MOVABLE CONTACT

THERMAL UNIT

LOAD CONNECTIONS
TO MOTOR TERMINALS

Figure 5-5.
Magnetic Starter.

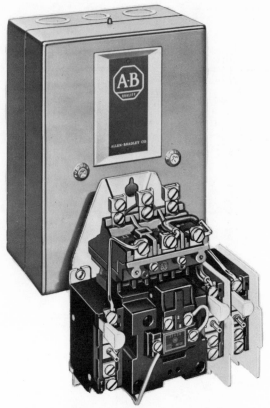

Courtesy of Allen-Bradley

Figure 5-6.

Magnetic Motor Starter with Nema Type 1 Enclosure.

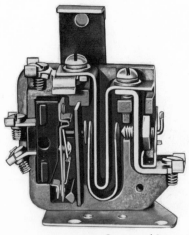

Courtesy of Square D

Figure 5-7.

Thermal Overload Relay.

a. *Courtesy of Square D*

b. *Courtesy of Square D*

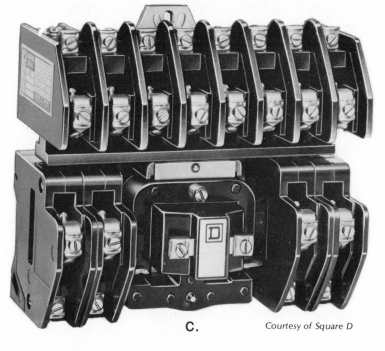

c. *Courtesy of Square D*

Figure 5-8.

Magnetic Relays and Contactors.

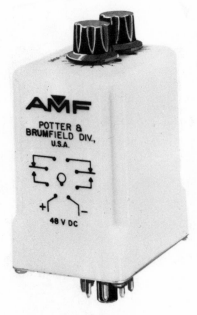

Courtesy of Potter & Brumfield

Figure 5-9.

Solid-State Relays and Time Delay Relays.

- Time-delay relays are either magnetic, pneumatic, or solid-state type. They have an added built-in mechanism to delay the opening or closing of contacts when a relay is energized. Figure 5-9 shows a dual solid-state time delay relay.

5. Solid-state and magnetic type multi-speed motor starter

   - Solid-state and magnetic multi-speed motor starters (usually two-speed) operate similarly to the preceding description. Their operation requires energizing one or the other contactor coils to select the desired motor speed. A time-delay device is normally used between the high and low speeds to avoid high ampere draw at initial low motor speeds. Figure 5-10 indicates a view of typical multi-speed motor starter within an enclosure.

6. Solid-state and magnetic type reversing motor starters

   - These are very similar to multi-speed starters, except when an alternate coil is energized in lieu of the initial contactor coil, the motor rotation is reversed. This is accomplished by changing two of the three phase legs supplying power to the motor. A time delay between rotation change is usually required to allow the motor to come to a full stop mode before it is reversed in rotation. A picture of a full-voltage reversing starter is depicted in an enclosure in Figure 5-11.

Courtesy of Allen-Bradley

Figure 5-10.

Multispeed Starters.

Courtesy of Allen-Bradley

Figure 5-11.

Full Voltage Reversing Starter.

7. Solid-state and magnetic type reduced voltage starters

   ■ Energizing a contactor in a motor starter, with its associated instantaneous closing, is referred to as an across-the-line starter. Often it is desirable to limit the maximum motor horsepower size for across-the-line starting use.

   ■ A reduced-voltage starter like an across-the-line starter energizes a coil that closes the power contacts to allow electricity to flow to the motor. However, a reduced-voltage starter through the use of an auto-transformer or similar ampere-limiting components can prevent an unwanted high initial current surge by reducing the input voltage to the motor being started.

   ■ Reduced voltage starters are frequently used for large motors to control the starting current (ampere) surge so that momentary voltage dips in the electric distribution system of the plant or building will be minimized.

   ■ Figure 5-12 contains views of solid-state reduced-voltage starters.

8. Combination Starters

   ■ When a solid-state or magnetic motor starter is incorporated into an enclosure with a circuit breaker, disconnect switch, or fused

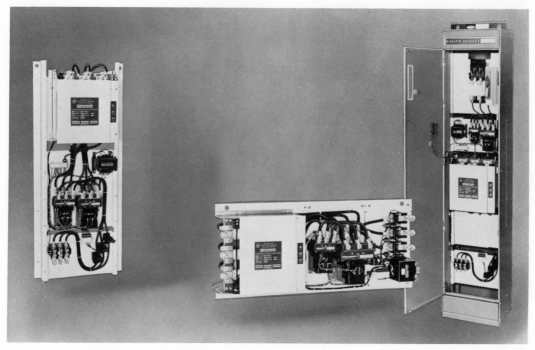

Courtesy of Allen-Bradley

Figure 5-12.

Solid-State Reduced Voltage Starter.

disconnect switch, it is termed a combination motor starter. The typical combination starters may be in the form shown in Figure 5-13. The reversing motor starter shown in Figure 5-14, as well as other motor starter types, can be arranged to be combination starters.

9. Motor Control Center

   ■ In some applications, it is more economical to connect many motors from a central power and control source. In lieu of mounting individual combination motor starters for each motor, a multiple modular package called a Motor Control Center can be used. A typical installation is shown in Figure 5-15.

10. Pilot Devices, Safeties and Accessories

   ■ An important component of the electric control system is the primary pilot (sensing or triggering) device. Also important are the safeties and accessories.

   a. In general, pilot devices are used to govern the on-off current flow in control circuits. This includes controlling relays, contactors, and starters. Pilot devices can take many forms, including: (1) pushbuttons, (2) thermostats, (3) float switches, (4) time clocks, (5) pressure switches, (6) mechanical and electrical alternators, etc.

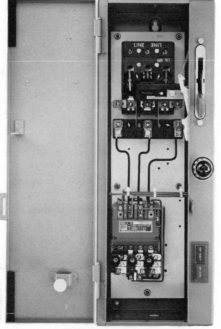

Courtesy of Furnas Electric Company

Figure 5-13.

Combination Starter Fused Disconnect-Type.

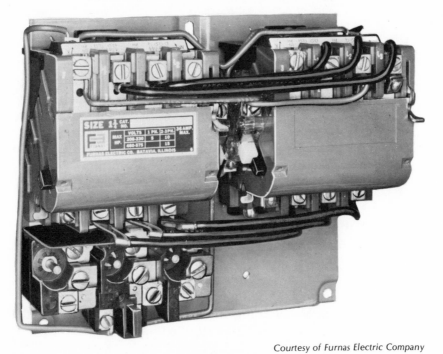

Courtesy of Furnas Electric Company

Figure 5-14.

Reversing Motor Starter.

Courtesy of General Electric Company

Figure 5-15.

Motor Control Centers.

b. A control device used to shut down a system and to prevent injury or damage to life and property is referred to as a *safety device*. This is usually accomplished by opening (stop) the control circuit. Some typical safety devices are: (1) high or low water cutout devices as used in boilers, (2) high-limit thermostats, (3) fire thermostats, (4) freeze thermostats, (5) low-limit thermostats, (6) smoke and fire detectors.

c. To visually indicate the on-off condition of equipment, pilot lights are frequently incorporated into control circuits.

d. Control transformers are used to change the line voltage power source (generally over 120 volts), to the equipment to be controlled, to a control voltage of 120 volts or less. In this manner pilot lights, safety devices, and other components are energized at 120 volts or less, thereby avoiding high voltage control circuits that require larger devices with heavy duty components.

Whenever the line voltage to the equipment to be controlled is 120 volts, the on-off devices (pilots) and safeties are often wired directly to the controlled equipment. This integrates the control and power circuits as one. However, when a three-phase power system supplies the controlled equipment, a 120-volt control circuit is used.

11. Multiplexing, Interfacing, and Programmable Controllers

a. System control and monitoring of remote components from a central location can involve significant numbers and lengths of cable. This could add appreciably to initial installation costs.

Multiplexing provides one method of avoiding large quantity of wiring. An example of this technique involves the hard wiring of several closely located or common control devices to a nearby input/output controller. From such a remote input/output controller, one pair of wires is installed to a centrally located programmable controller (PC) or computer.

The single pair of wires carries pulsating electric signals that identify an individual control device. These signals are generated and interpreted by the programmable controller (PC), and/or computer. This method enables many remote devices to be controlled simultaneously via a minimum number of wires.

b. The combining of two distinct systems to operate as one is called interfacing. When controlled devices are hard wired to an input/output controller for transmission to the PC or computer, they are termed as *interfaced* into the PC system. Similarly, when a pneumatic system is activated by opening an electrical solenoid air valve, interfacing occurs.

LADDER DIAGRAMS

The logic for understanding ladder (elementary) diagrams is explained in this section. The technique used includes an explanation of the construction and logic of 12 diagrams. This procedure starts with the simplest of preliminary diagrams and progresses to more complex ladder diagrams by adding additional features. In the simpler diagrams components are enclosed in square or rectangular outline boxes for easier understanding.

Single-Phase Line Voltage Control

Figure 5-16 shows a simple wiring diagram (5-16a) and the corresponding ladder diagram (5-16b) for a single-phase 120 volt motor control circuit. The wiring diagram identifies the 120 volt supply, a control thermostat in the "hot leg", and a manual motor starter near the agitator motor AG-1. This starter consists of a manual switch and a thermal motor overload that protects the agitator motor AG-1.

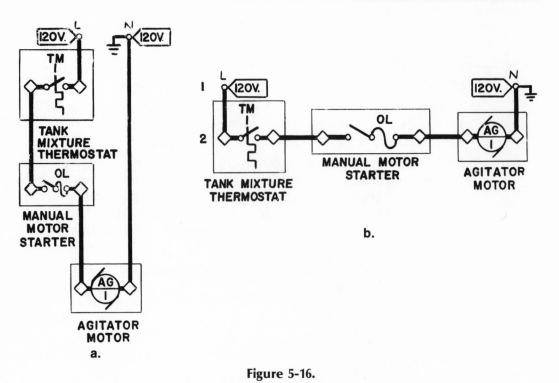

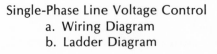

Figure 5-16.

Single-Phase Line Voltage Control
a. Wiring Diagram
b. Ladder Diagram

With this wiring diagram, an electrician would have no difficulty wiring to the system components. The ladder diagram depicts the logic involved when the thermostat, which is the primary controller in this example, closes the circuit. If the manual switch is closed, the agitator motor AG-1 will start. Either the manual switch or the thermostat can open the circuit and stop the motor. Of course, the thermal motor overload can also interrupt the circuit.

Single-Phase Pilot Control

Figure 5-17 shows a control condition similar to that described for Figure 5-16, except the control circuit is reduced from a 120 volt supply to a 24 volt control voltage. The wiring diagram (5-17a) shows each element in its respective position for effective wiring. The five-step ladder diagram (5-17b) shows the logic of the control circuit via the rungs:

1. The power source is 120 volts.

2. It supplies the high voltage side of the transformer in the control circuit.

3. The low voltage side of the transformer supplies the control voltage at 24 volts. This pilot control voltage is more sensitive than higher voltage and

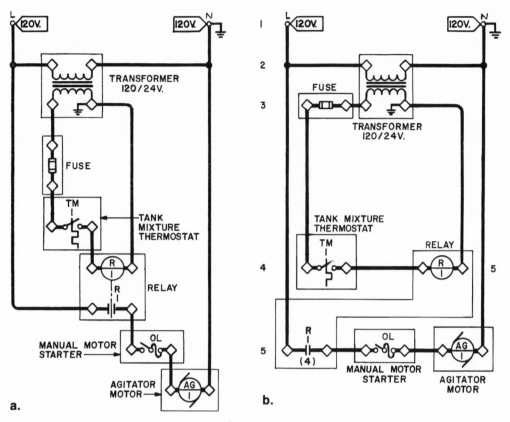

Figure 5-17.

Single-Phase Pilot Control
a. Wiring Diagram
b. Ladder Diagram

therefore controllers are easier to fine adjust. This circuit is grounded and a fuse protects the 24 volts control circuit from overload.

4. On Rung 4 the thermostat is the main controller. When the thermostat closes, the relay R-1 in the control circuit is energized. The notation "5" to the right of the fourth rung indicates Rung 5 contains a device that is activated when the relay R-1 is energized.

5. When this occurs, the contacts R-1 in the hot leg of the electrical supply to the motor are closed. If the switch in the manual motor starter is closed, the agitator motor AG-1 will start. The motor will continue to operate until the circuit in Rung 5 is interrupted. This can be accomplished by opening the manual switch in the motor starter, a motor overload triggering the thermal overload device, or when the relay contacts are opened. The contacts are opened only when the 24 volt control circuit is interrupted, usually by the thermostat; but they can also be opened by a short circuit or an overload causing the fuse to open the control circuit. Note: The "4" under contacts R-1 indicates this device is controlled by an element in Rung 4.

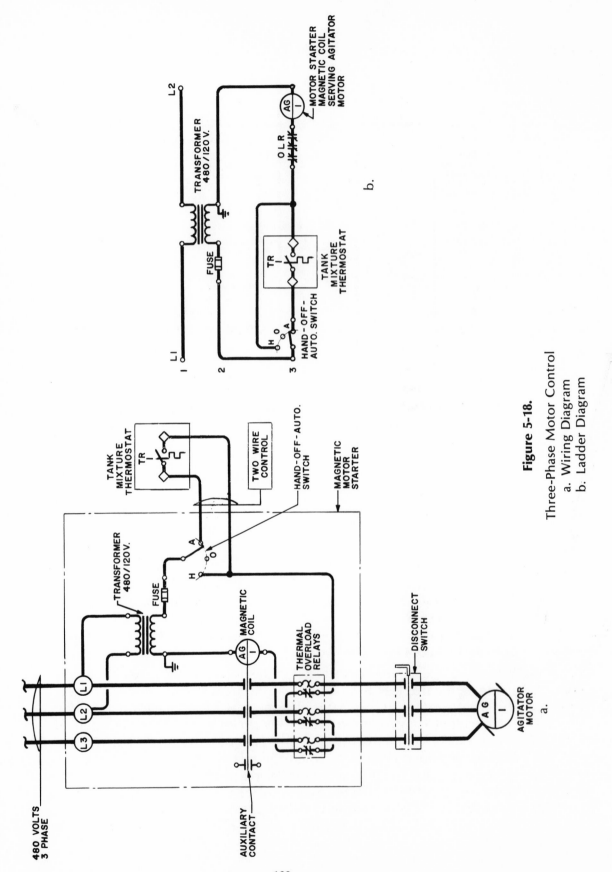

Figure 5-18.

Three-Phase Motor Control
a. Wiring Diagram
b. Ladder Diagram

160

Three-Phase Motor Control

Figure 5-18 shows the wiring and ladder diagrams for a motor control circuit which is similar to Figures 5-16 and 5-17. The major difference in this example is the three-phase 480 volt supply to the agitator motor. The wiring diagram (5-18a) shows all components and connections in both the 480 volt supply and the 120 volt control circuit. The ladder diagram (5-18b) shows only the elements in the control circuit.

For example, the manual disconnect switch at the motor, normally used as a maintenance safety feature, is shown in the wiring diagram. It is excluded from the ladder diagram, however, because it has nothing to do with the control circuit. In the four-step ladder diagram, all aspects of the control circuit are shown in their logical sequence, as follows:

1. Line voltage at 480 volts is taken from any two hot leads of the three-phase, three-wire supply.

2. The high voltage (480 volts) side of the transformer is reduced to the control circuit low voltage (120 volts) side, where the hot leg has a protective fuse and the neutral side is grounded. This could be reduced to 24 volts or any other control voltage.

3. The interlocking controls are in series. They include a three-position "hand-off-automatic" selector switch, a thermostat that is effective only when the switch is on automatic, and thermal overload relay devices to protect the agitator motor AG-1. With this type of control circuit, the motor after power restoration will start up unattended; provided the three-way switch (H,O,A) is in the automatic mode and the thermostat closes the circuit, or if the three-position switch is in the "hand" mode.

   This type of control is commonly known as a two-wire control circuit because two wires are required to connect the thermostat to the circuit. *Note:* the magnetic coil and motor are both identified as AG-1, but their symbols are slightly different. In addition, the wiring diagram shows auxiliary contacts that are used to control auxiliary equipment. The use of the auxiliary contacts is demonstrated in the next ladder diagram.

Manual Three-Phase Control

Figure 5-19 contains a ladder diagram that is similar to Figure 5-18, except a manual reset is required after a power failure. This motor control circuit is commonly referred to as a three-wire control circuit because three wires are required to connect the remote stop-start pushbutton station feature to the circuit. Also, auxiliary contacts, shown in the wiring diagram, are used to supply constant control voltage to the control circuit. The five-rung ladder diagram contains the following logic:

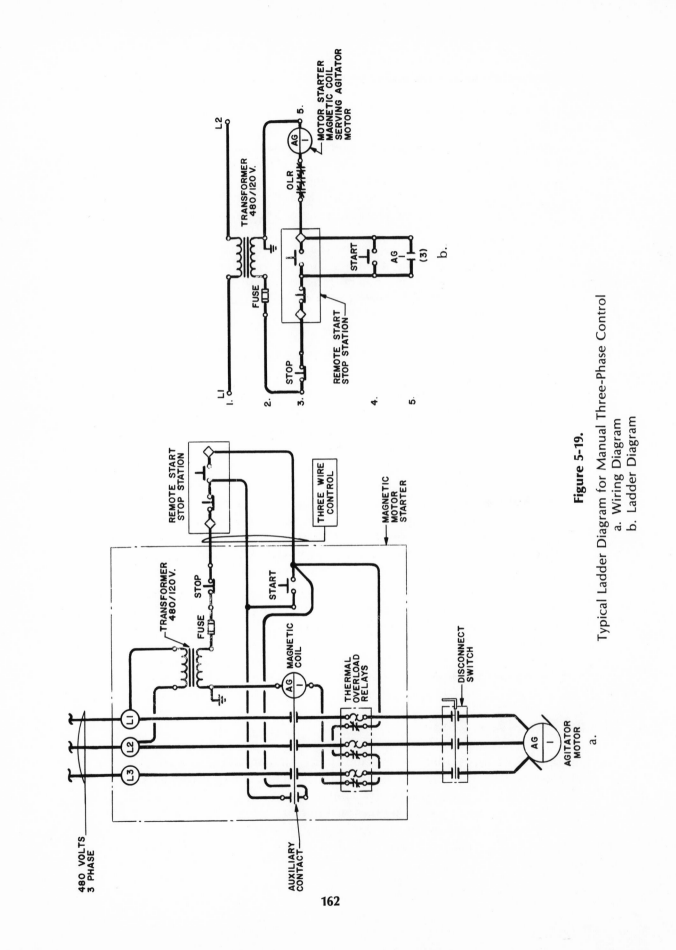

Figure 5-19.

Typical Ladder Diagram for Manual Three-Phase Control

a. Wiring Diagram
b. Ladder Diagram

162

1. Line voltage is supplied to the high voltage side of the transformer at 480 volts.

2. The 120 volts control voltage has a fuse in the hot leg and is grounded in the neutral leg.

3. The two stop pushbuttons, one at a remote location, are in series with one another and both are normally closed. The start pushbuttons are normally open. One start pushbutton must be manually (momentary push) closed to energize the motor. The thermal motor overloads, which protect the motor, are normally closed. When the magnetic coil AG-1 is energized, it closes the main motor power contacts and simultaneously the auxiliary contacts AG-1 also close.

4. The start pushbutton that is remotely located is parallel to the local start pushbutton and with the auxiliary contacts AG-1.

5. When the magnetic coil energizes auxiliary contact AG-1, which is parallel to the two parallel-connected start pushbuttons, the motor control circuit will remain closed when the manual start pushbutton is released. However, if a power failure occurs, the magnetic coil AG-1 will be de-energized and the entire control circuit will open. When power is restored, the motor will not start until a start pushbutton is manually closed.

Single-Phase Automatic Start-Up Control

Figure 5-20 is very similar to the control wiring diagram in the previous example, except controls are in parallel in Figure 5-19, and in Figure 5-20 they are in series. Accordingly, this is a two-wire circuit containing the automatic start-up feature following a power failure. *Note:* wiring diagrams are excluded from the remaining figures because ladder diagrams are the primary subject of this chapter. The control features in this three-step ladder diagram are described as follows:

1. The 120 volt power source is attainable directly from the line service; or from a transformer.

2. An aquastat is normally closed, but will interrupt the circuit when the water temperature in the pipe is below a predetermined level. The three-position selector switch, shown in the automatic position, also has off and hand (manual) positions. The hand position enables the circuit to bypass the main controller, which is a room thermostat. The thermostat is normally open and closes the circuit to engage the unit heater motor. Finally, the manual motor starter contains the typical switch (manual, normally closed except for maintenance) and the thermal motor overload.

3. A green pilot light is in parallel with the motor, when the motor is operating, the light is on.

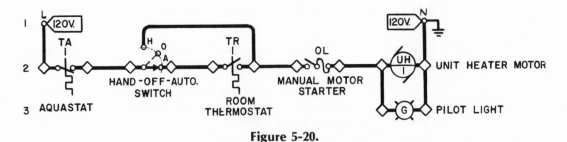

Figure 5-20.

Ladder Diagram for Single-Phase Automatic Start-Up Control.

Automatic Control of Single-Phase Exhaust Fan

Figure 5-21 contains the ladder diagram for controlling a typical single-phase 120-volt exhaust fan. The control circuit includes several safety features as noted in the following ladder diagram explanation.

1. The 120-volts power source supplies the control circuit.
2. This is the key control rung in the ladder diagram, and it contains the following elements.
   - A room-type low-limit thermostat TL-1 is normally closed. When the room temperature drops below 55°F, it will open, thereby interrupting power to the fan.
   - The fire thermostat TF-1 is located in the exhaust fan suction plenum. It is normally closed, but will open at the high temperature of 125°F, which can be produced by a fire. If smoke detection and control is required, a fire and smoke detector would be substituted for the fire thermostat and wired in series on this rung.
   - The three-position selector switch is shown on automatic, but the hand position H enables the circuit to bypass the main controller TR-1.
   - A room-type thermostat TR-1 automatically starts the fan when the room temperature reaches a predetermined temperature. At that point, the circuit is closed and the fan starts up. When a preset lower temperature level is reached, the thermostat opens the circuit stopping the fan.
   - The manual motor starter is depicted by a manual switch and thermal motor overload. This manual switch is normally in the closed position and primarily used for maintenance purposes.
   - Finally, a fan motor is shown on this 120 volt single-phase power and control voltage line.
3. A green lamp illuminates when the motor is operating.
4. The normally closed exhaust damper DE-1 is opened when the fan is operating. The damper motor is equipped with a manual motor starter

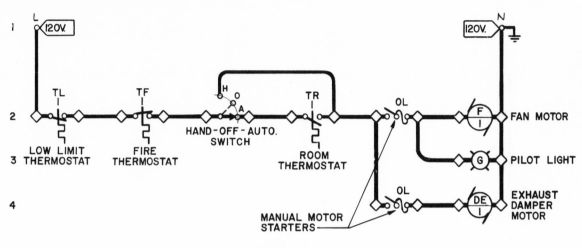

Figure-5-21.

Ladder Diagram for Automatic Control of Single-Phase Exhaust Fan.

(switch and thermal motor overload). The manual switch is normally closed.

In this two-wire control circuit the fan will start up automatically following a power failure, provided the room temperature is higher than the room thermostat setting.

Manual Control of Three-Phase Exhaust Fan

Figure 5-22 exhaust fan motor control is the same as Figure 5-21 except for manual control (start-stop) and in the use of a three-phase fan motor.

1. Power is taken from two hot legs of the 480 volt supply. Therefore, the high voltage side of the transformer is at line voltage.

2. The transformer secondary provides control voltage at 120 volts. The hot leg of the control circuit is fused to protect the components, and the neutral leg is grounded.

3. The third rung of the ladder diagram contains the principal controller and safety devices. These operate as follows:

   ■ The normally closed low-limit thermostat TL-1 will open the circuit when the room temperature drops to 55°F.

   ■ The fire thermostat TF-1 is located in the exhaust fan suction plenum. It is normally closed, but will open at the high temperature of 125°F, which can be produced by a fire. If smoke detection and control is required, a fire and smoke detector would be substituted for the fire thermostat and wired in series on this rung.

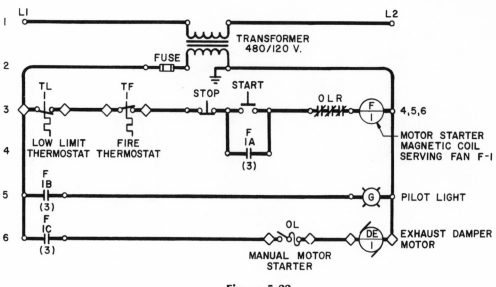

Figure 5-22.

Ladder Diagram for Manual Control of Three-Phase Exhaust Fan.

- The stop pushbutton is normally closed. However, when depressed, the circuit is opened, causing the exhaust fan to stop.

- The start pushbutton is normally open. When closed, the circuit is complete and the fan motor starts. However, when the start pushbutton is released, the switch opens that part of the circuit. Therefore, the maintaining contact in Rung 4 is necessary to keep the fan motor operating. This is further explained below:

- The motor starter thermal overload relays are normally closed, and are disengaged only when line current to the motor is too high. At overload conditions, these contacts open the control circuit and the fan motor stops.

- The fan motor magnetic contactor F-1 is energized when the third rung circuit is closed. When energized, this coil closes the line voltage contacts to the motor, thereby starting it. The magnetic contactors also close the auxiliary contacts shown on Rungs 4, 5, and 6.

4. When Rung 3 circuit is closed, the contact F-1A is closed and remains closed until the fan circuit is interrupted. The purpose of the contact is to lock in the start pushbutton and enable the fan motor to continue operating (circuit remains closed) after the start pushbutton is released. Once the control circuit on Rung 3 is interrupted, auxiliary contact F-1A is returned to its normally open position. The motor will not restart until the start pushbutton is depressed.

5. Auxiliary contactor F-1B is normally open. However, when the fan operates, it is closed and the green pilot light is illuminated.

6. Auxiliary contactor F-1C closes the circuit in Rung 6, when the fan operates. As a result, the exhaust damper motor is energized and the damper opens. When the fan shuts down, auxiliary contactor F-1C is opened and the damper is closed. A manual switch and thermal motor overload, both normally closed, are located near the damper operating device.

Automatic Control of Three-Phase Exhaust Fan

Figure 5-23: The control circuit for this exhaust fan motor is similar to Figure 5-22, except it has an automatic control feature. This means the room thermostat is the prime control when the three-position selector switch is in the automatic position.

1. The line voltage is supplied to the primary high voltage side of the control circuit transformer.

2. The control voltage is at 120 volts. The hot leg has a lockout switch, which, although not an approved disconnect means, is a good maintenance feature. Rung 2 also has a circuit fuse. The neutral side is grounded.

3. Rung 3 of the ladder diagram contains the principal controller and safety devices. These operate as follows:

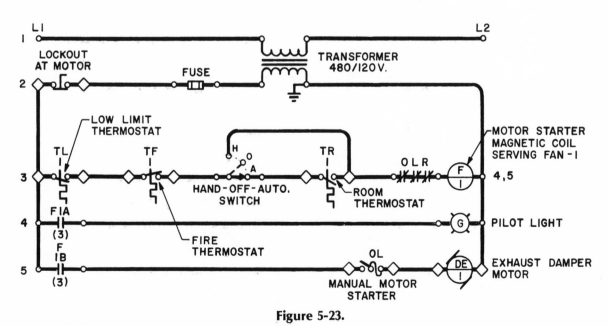

Figure 5-23.

Ladder Diagram for Automatic Control of Three-Phase Exhaust Fan.

- The low-limit thermostat TL-1, the fire thermostat TF-1, the motor starter thermal overload relays OLR, and the magnetic coil F-1 operate the same as outlined in the Figure 5-22 description (see Rung 3). The three-position selector switch and the room high-limit thermostat TR-1 operate the same as outlined in the Figure 5-21 description (see Rung 2).

4. The normally open auxiliary contacts F1A and F1B are closed whenever the exhaust fan motor starter magnetic coil is energized. The contacts operate the same as explained for Figure 5-22.

Mechanical Alternation of Two Single-Phase Motors

Figure 5-24 shows the control circuit ladder diagram for a single-phase mechanical alternator that is alternating the use of two motors. This is typical for controlling the fluid level in a vessel with two independently driven single-phase pump motors. The control circuit logic for single-phase motors is explained as follows:

1. The 120 volts power source is grounded on the neutral leg and has a fuse for circuit protection in the hot leg (not shown).

2. The mechanical alternator contacts are normally open. They are closed only when the alternator selects pump motor P-1 for operation. The

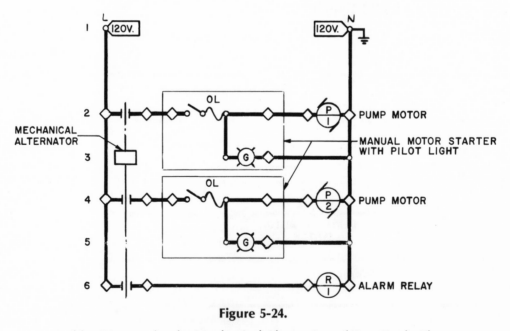

Figure 5-24.

Ladder Diagram for the Mechanical Alternation of Two Single-Phase Motors.

manual motor starter switch and thermal motor overload OL are normally closed. The motor P-1 is energized when the circuit is closed and the motor operates the pump. When the alternator contacts are opened, the motor stops running.

3. The green indicator light is on when motor P-1 is operating.

4. The mechanical alternator (Square D, Type AW, Class 9038 is a representative float-controlled alternator shown in Figure 5-26) is a mechanical device that alternates the use of two motors, each capable of meeting operating requirements. When the float device (presumably in a tank) reaches a primary level, the alternator calls for motor P-1 to operate by closing the contacts in Rung 2. Pumping is completed when the float lowers to a predetermined lower level and the alternator disengages the contacts in Rung 2. The next time the float reaches the primary control level, the alternator closes the contacts in Rung 4, thereby energizing pump motor P-2. This cycle is repeated sequentially, alternating the use of and wear on the two-pump motors.

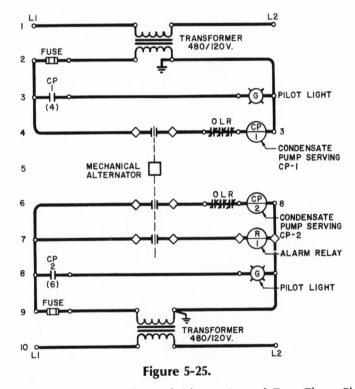

Figure 5-25.

Ladder Diagram for Mechanical Alternation of Two Three-Phase Motors.

The alternator also has two other features. First, when the water rises to a secondary control level (above the primary level), the alternator closes the contacts on both Rung 2 and Rung 4, thereby operating both pumps. This higher level would be reached when the inflow to the tank exceeds the capacity of a single pump. Second, if the water level continues to rise above the secondary level the alternator will then close another set of contacts shown on Rung 6 and an alarm will sound. This will occur when the inflow exceeds the capacity of both pumps or a malfunction of the pumps.

5. Rungs 4 and 5 are controlled exactly as explained for Rung 2 and Rung 3.

6. As explained in Rung 4, when a high level is reached, the contacts will be closed by the alternator and alarm R1 will be energized. This will signal the imminent overflow of the tank.

Mechanical Alternation of Two Three-Phase Motors

Figure 5-25: This control circuit is similar to Figure 5-24 except the alternator, which is the primary controller, is controlling two condensate pump motors each powered by three-phase current. The logic of each rung in the ladder diagram is as follows:

1. The line voltage is supplied from two of the three-phase feeding condensate pump motor CP-1. This line voltage supplies the primary high voltage side of the transformer.

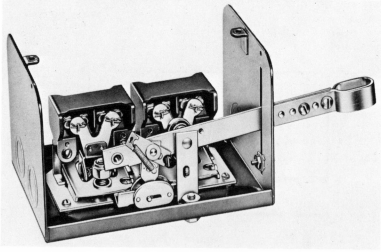

Courtesy of Square D

Figure 5-26.

Mechanical Alternator.

2. The low voltage side of the transformer is at 120 volts, which is the control circuit voltage. The hot leg of the circuit is fused and the neutral side is grounded.

3. Auxiliary contact CP-1 is normally open, but when magnetic coil CP-1 is energized, this coil is closed. The circuit is complete, and the green pilot light lamp illuminates, indicating motor CP-1 is operating.

4. This is the main control rung for motor CP-1. When the mechanical alternator closes the contacts in this rung, the circuit is completed. Magnetic coil CP-1 is energized and the motor supply line voltage contacts are closed. The motor starts and continues until the mechanical alternator opens the contacts on this rung.

5. The operation of the alternator is exactly the same as described at Rung 3, Figure 5-24.

6. This rung functions exactly the same as Rung 4 above; except this rung controls condensate pump motor CP-2.

7. As explained for Figure 5-24, Rung 6, an alarm R1 is sounded when the liquid level in the tank reaches a very high level. This circuit is activated when the alternator closes the contacts in this rung.

8. This rung operates the green pilot light the same as Rung 3 above.

9. & 10. The two rungs perform the same function for motor CP-2 as Rung 1 and Rung 2 perform for motor CP-1. Each motor has its own line voltage supply to its control circuit. With this arrangement, each motor (and its control) is completely independent from the other. This is advantageous for maintenance and emergency repairs.

Manual Control

Figure 5-27: This ladder diagram identifies the control logic for a manually operated supply air unit powered by three-phase current.

1. The line voltage is taken from two phases of the three-phase power supply to the motor. The line voltage input to the control circuit is through the primary high voltage side of the transformer.

2. The 120 volt control circuit is fused and grounded.

3. This is the main control circuit that is protected by a normally closed smoke detector SD-1 and a normally closed freeze thermostat TZ-1 in Rung 6. The start-stop switches, the thermal motor overload relays OLR, and the motor magnetic starter F-1 are operated exactly the same as described for Rung 3 in Figure 5-19. The magnetic coil F-1 operates auxiliary contacts in Rungs 5, 7 and 9.

4. By its parallel arrangement with the local start button the remote start pushbutton can energize magnetic coil F-1.

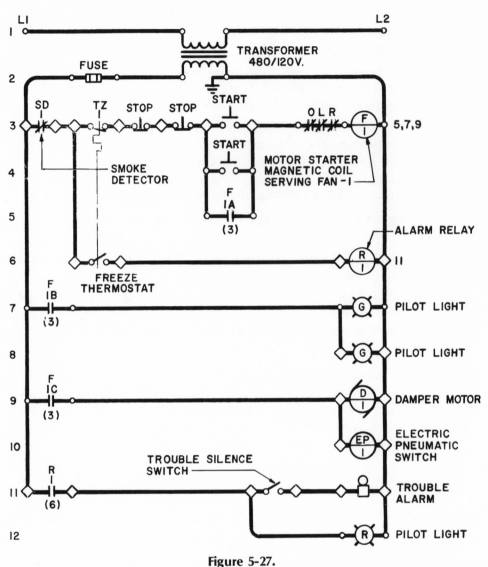

Figure 5-27.

Ladder Diagram for Manual Control.

5. Auxiliary contact F-1A is energized by magnetic coil F-1. It remains closed while the motor operates. When the motor stops (Rung 3 circuit is interrupted) coil F-1 is de-energized and contacts F-1A are opened (the normal position). The motor cannot be restarted until a start pushbutton is engaged.

6. The freeze thermostat is located in the fresh air supply duct. When the inlet temperature (downstream of the heating coil) drops to 36°F, the freeze thermostat opens the contacts in Rung 3. It also energizes relay coil R1. The TZ-1 control device opens the main control circuit, and the air

supply fan motor stops (if operating) and will not restart until the condition is corrected. The relay coil R1 closes contacts in an alarm circuit described below.

7. Auxiliary contact F1B is closed when magnetic coil F-1 is energized. This closes the circuits in Rungs 7 and 8. A green pilot light is illuminated at the motor starter.

8. When contacts F-1B are closed, a remote green pilot light will also illuminate. Both lights indicate the air supply fan motor is running.

9. & 10. When magnetic coil F-1 is energized, the auxiliary contact F-1C is closed and the fresh air intake dampers are opened. These are usually opened by electric damper motors D-1 or an electric pneumatic switch.

11. When relay coil R1 is activated by the freeze thermostat in Rung 6, the contacts at R1 are closed and the circuits in Rungs 11 and 12 are completed. The alarm sounds and a manual switch is available to silence the audio alarm.

12. The red pilot light alarm remains on as long as the contacts at R1 Rung 11 are closed. The red pilot light will not go off until the freeze thermostat TZ is above 36°F and reset.

Automatic Control of Three-Phase Supply Air Unit

Figure 5-28 is a ladder diagram very similar to Figure 5-27, except this one is automatically controlled. It also includes a fire thermostat TF safety device.

1. & 2. These rungs are exactly the same as the explanation in Figure 5-27.

3. The main control circuit has the following features:

- A freeze thermostat and a fire thermostat are two safety devices in this primary control circuit. They are normally closed, but they interrupt the circuit when either set of contacts are opened by their respective thermostat.

- The three-position selector switch (hand, off, automatic) is usually in the automatic position. The manual position (sometimes called the hand position) bypasses the primary control, which is called the pilot device.

- For a supply air system, the pilot device would normally be a thermostat, but in some process operations it could be a time clock or similar control device. A time clock is shown in this example.

- The thermal overload relays OLR and fan motor coil F-1 are the same to those described in previous figures. In this application the magnetic coil F-1, when energized, closes the auxiliary contacts in F-1A and F-1B on Rungs 6 and 8, respectively.

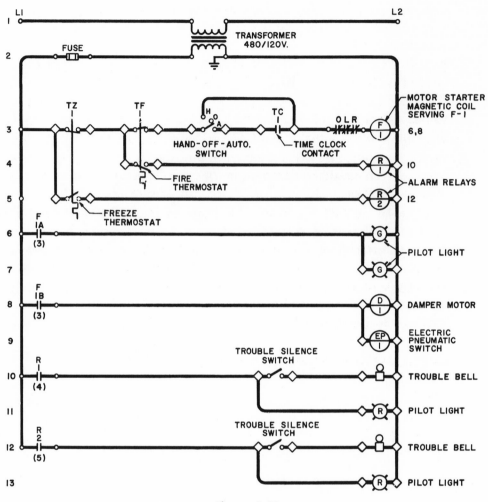

Figure 5-28.

Ladder Diagram for Automatic Control of Three-Phase Supply Air Unit.

4. The fire thermostat functions similarly to the freeze thermostat except for its operating temperature and location. Fire thermostats are usually activated at about 125°F. When activated, the normally closed contact in Rung 3 would be opened. In this application, the fire thermostat would probably be located in the inlet duct to the fan.

5. This freeze thermostat rung is exactly the same as described in Rung 6 Figure 5-27. When the freeze thermostat is closed, relay coil R2 will also be activated. This relay controls the alarm circuits in Rungs 12 and 13.

6. & 7. When magnetic coil F-1 is energized, the auxiliary contact at F-1A is closed. This completes the circuit and green pilot lights in both rungs

signal that the motor is operating. One of the lights is located at the motor starter. The other pilot light is remotely located (possibly in a master control panel).

8. & 9. When magnetic coil F-1 is energized, the auxiliary contacts at F-1B are also closed. This completes the circuits in Rungs 8 and 9 and the air inlet dampers are opened. These are usually actuated by electric motors or an electric pneumatic switch. When coil F-1 is de-energized, the contacts at F-1B are opened and the dampers closed by a spring drive.

10. When relay coil R1 in the freeze thermostat Rung 4 is energized the contacts at R1 are closed. This completes the circuits in Rungs 10 and 11. The audio alarm (bell) sounds, indicating the presence of cold air in the intake duct. A silencing switch is located in this circuit to silence the audio portion of the alarm.

11. The red trouble pilot light also goes on when the audio alarm sounds. However, when the silencing switch turns off the audio alarm, the trouble light remains on until the air temperature in the intake duct is increased to above 35°F and the freeze thermostat is reset.

12. The contacts at R2 are closed when the relay coil R2 in the fire thermostat Rung 5 is energized. This completes the circuits in Rungs 12 and 13. The audio alarm sounds, indicating the presence of hot air (over 125°F) in the outlet duct. A silencing switch is located in this circuit to turn off the audio alarm.

13. The red trouble pilot light also goes on when the alarm bell sounds. However, when the silencing switch turns off the alarm, the trouble light remains on until the discharge air temperature is decreased below 125°F and the fire thermostat is reset.

APPLICATION

Background of a Representative Complex Ladder Diagram

The confidence gained by understanding the 12 ladder diagrams in the previous section provides the maintenance trouble-shooter with the analyzing skill necessary to take on control circuit maintenance. However, control circuits may be electrical, pneumatic, electronic, solid-state, or a combination of these elements. Accordingly, the typical application explained in this section includes all of these control elements. As mentioned before, the logic is the same regardless of the control medium.

Figure 5-29 contains a floor plan and an air flow diagram of a complex heating-ventilating system for three laboratory rooms. Each room has its complete and independent air handling system with corresponding discharge ducts and fans. The heating-ventilating (HV) units for Room No. 1 are electrically controlled. In Room

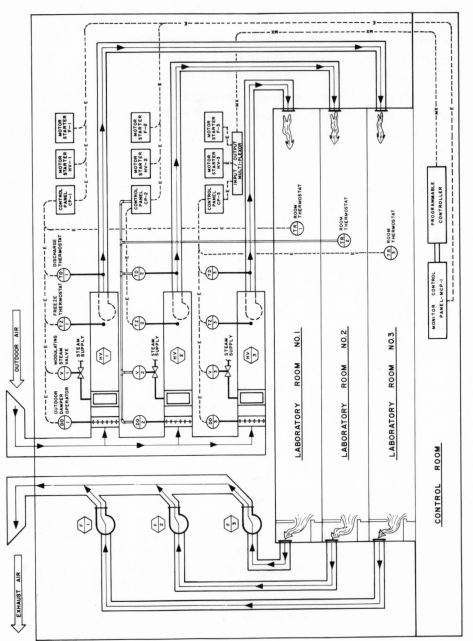

Figure 5-29.

Air Flow Diagram for Heating and Ventilating Three Laboratories.

No. 2, the HV control is electric/pneumatic; and, in Room No. 3, the HV controls are electric/solid state. This third system contains remote control capability using a programmable controller (PC) centrally located in a control room. It is made more complex by multiplexing the signals between the PC and the remote local control stations.

Consolidating these three different control mediums into one system is complicated and difficult to comprehend without a ladder (logic) diagram. However, with a properly prepared ladder diagram, and a meticulous step-by-step analysis, the control system logic can be understood. Figure 5-30 provides a ladder diagram for this example.

The structure of the ladder diagram depicted in Figure 5-30 has been designed for ease of following the logic. Each of the four vertical groupings apply to a major segment of the control system, as follows:

1. Rungs 1 through 17 identify those circuits that are centrally located in monitor control panel CP-1 for systems HV-1, HV-2 and HV-3. (Note: The HV-3 system is interfaced to the PC.)

2. The second vertical ladder groupings, Rungs 18 to 34, contains the logic for the electrical control of system HV-1.

3. The third ladder grouping, Rungs 35 to 52, identifies the logic of the electrical/pneumatic control circuit for system HV-2 serving the heating ventilating system for laboratory room No. 2.

4. Finally, the fourth grouping, of this ladder diagram, Rungs 53 to 70, contains the logic circuitry in the electric/solid-state used in conjunction with the PC.

The following explanation of this complex control system will parallel the analysis of the four-part ladder diagram. However, first a clarification of minor variations from ladder diagram explanations used in the previous section is in order. These include the following:

1. Selected rungs in this final ladder are left open to provide separation between key component segments in the logic.

2. Rungs in the four vertical ladders diagram groupings are conveniently numbered for ease of identification and explanation.

3. In this complex ladder diagram example, following the logic requires a path that varies from the rung numbering sequence commonly used.

4. For convenience and brevity, symbols and terms that are explained earlier are not repeated in the following ladder diagram explanation.

Explanation of This Representative Ladder Diagram

1. Rung 1 indicates the central control circuit is powered at 120 volts.

2. Rung 2 contains the 120 volt power supply to drive the program clock.

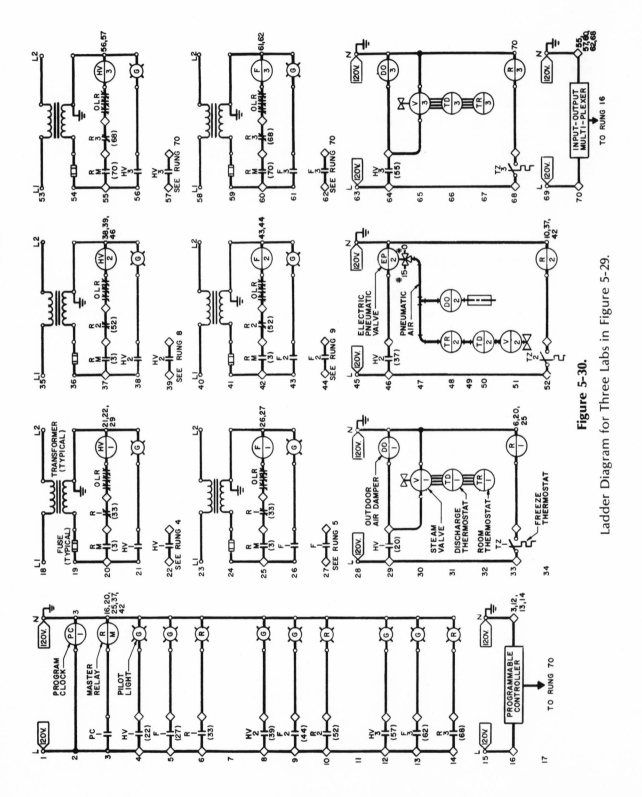

Figure 5-30. Ladder Diagram for Three Labs in Figure 5-29.

3. When the program clock contact PC-1 is closed, the Master Relay RM is energized. This action closes contacts in Rungs 20, 25, 37 and 42. It also sends an impulse to Rung 16. In following the logic, the effect of RM on the second vertical ladder grouping at Rungs 20 and 25 are reviewed first. Subsequently, the third and fourth vertical groupings are analyzed in their turn.

4. In Rung 20, RM contactors are closed and the entire circuit on this rung is energized. R-1 is an override described in the Rung 33 explanation. The overload relays open the circuit whenever the motor overheats. When magnetic coil HV-1 is energized, contactors are closed in Rungs 21, 22 and 29. Also, the magnetic motor starter contactors are closed starting the motor which drives the heating-ventilating fan for laboratory room No. 1 (system HV-1).

5. In Rung 21, the closing of these contactors turns on a green pilot light located near the magnetic motor starter.

6. Rung 22 indicates HV-1 in Rung 4 closed the circuit, thereby lighting a green pilot light at the central control station. These green pilot lights indicate that fan motor HV-1 is operating.

7. Rungs 18 and 19 identify control circuit powered from two legs of the HV-1 motor power supply source. This power is stepped-down to a control voltage (usually 120 volts) that is properly fused and grounded.

8. Rungs 29 to 32 contain the heating coil control circuit. The heating coil is located in the outside air intake duct. When magnetic coil HV-1 Rung 20 is energized, the magnetic coil contactors HV-1 in Rung 29 are closed. This energizes the circuit and the outside dampers DO-1 are opened. Simultaneously, the steam valve V-1 is energized and capable of modulating to a controlled position. The valve modulated position is electrically-controlled by two thermostats. TR-1 is the room thermostat in room No. 1; and TD-1 is the thermostat in air discharge duct of system HV-1.

9. Rungs 23 and 24 perform the same power source control for Exhaust Fan F-1 as described for motor HV-1.

10. Rung 25 is also energized when the master relay RM closes the contactor RM. The contacts at R1, controlled by Rung 33, and the motor overload relays OLR can interrupt this circuit. The energized coil at F1 closes the contactors on the fan motor starter and also on Rungs 26 and 27.

11. When contactor F-1 is closed on Rung 26, the local green pilot light is illuminated indicating the fan motor is operating.

12. Rung 27 identifies the contacts that are also at Rung 5. When closed, a green pilot light at the central control panel indicates the exhaust fan F-1 is operating.

13. Rung 33 contains the freeze thermostat located in the inlet air duct. When

inlet air is too cold, the freeze thermostat closes and relay coil R1 is energized. This action causes the contacts in Rungs 20 and 25 to open, thereby stopping both motors. Simultaneously, the contacts R1 in Rung 6 are closed and a red warning pilot light is illuminated on the control panel.

14. Rung 28 shows power supply at 120 volts controls the freeze thermostat circuit Rung 33 and the steam coil circuit Rungs 29 to 32.

15. The control logic for system HV-1 is explained in the previous 14 steps. The discussion continues by returning to Rung 3 and explaining control logic when the master relay RM closes the contacts at Rungs 37 and 42. These rungs are located in the third vertical ladder grouping, which is an electrical/pneumatic control system.

16. The control logic in Rungs 35 to 45, Rung 52 and Rungs 8, 9, and 10 is identical to the explanations for Rungs 18 to 28, Rung 33 and Rungs 4, 5, and 6.

17. The interface between electric and pneumatic control exists at Rungs 46 and 47. When the system HV-2 magnetic motor starter coil HV-2 on Rung 37 is energized, auxiliary contactor HV-2 closes the circuit on Rung 46. The three-way electric/pneumatic valve EP-2 is energized and the 15 psi control air supply enters the pneumatic control system. When the coil is de-energized, the valve is closed and the pneumatic system is opened to atmosphere. The three-way pneumatic valve is normally opened to the atmosphere. It closes when coil EP-2 is energized. At that moment, the 15 psi control air supply enters the pneumatic control system.

18. At Rung 48, the outside damper DO-2 is pneumatically operated and the intake dampers are opened.

19. The steam valve V-2 modulates the steam flow to the heating coils. However, this valve is controlled by pneumatic thermostats TR-2 and TD-2 shown on Rungs 48 and 50.

20. As stated above, the all-electric and electric/pneumatic systems perform the same functions and the logic is identical. Only the conveying medium is different. In the fourth and final ladder diagram grouping of this complex control system (electric/electronic) the conveying medium (electricity) is the same as the all-electric control circuit, but two modern technological improvements are introduced. These are the programmable controller and multiplexing. It is the intent here only to introduce you to these two control devices.

21. The explanation of control logic for the fourth vertical ladder, grouping the electric/electronic control, once again returns to Rung 3. Here, the master relay RM transmits a signal to the PC located on Rung 16. The PC in turn, analyzes the signal, determines what should be done, and transmits an appropriate signal by multiplexing to the input-output (I-U)

multiplexer at Rung 70. The multiplexer directs the signal to the proper rung in this fourth vertical ladder grouping. This procedure is also reversed from component to multiplexer to PC to the proper rung in the control center in the first vertical ladder grouping. There are many advantages to these types of electronic controls, but the more important ones involve the following:

- Multiplexing enables many different signals to be transmitted long distances (from central control to remote station) over two wires. This is accomplished through variations in signal impulse and intensity.

- PC systems enable remote components to be controlled automatically from a central location. They also permit a limited number of people to control an almost limitless number of control devices in many control systems. Finally, PCs have the capability of rearranging a control circuit from a control panel key board, hence the term *programmable*.

22. The logic of control for Rungs 53 to 68 and Rungs 12 to 14 are identical to control for Rungs 18 to 33, and Rungs 4 to 6, with two exceptions. These are the multiplexing between the first and fourth vertical ladder groupings, and the PC centrally located in a control room.

This representative example of a system ladder diagram highlights the following salient points to the maintenance operator:

1. The basic start-stop control and actuating of control systems is no different in electric, electronic, solid-state, or pneumatic systems.

2. The description of sophisticated modulating controls can be shown in a ladder form and maintain the system logic.

3. A ladder type diagram can be used to describe and understand a programmable controller or a computer. Their logic is the same. Their components may disguise it from the basic hard wire concept. However, the ladder diagram sequentially links them together.

The use and understanding of control systems with ladder diagram logic is an important tool that will not become obsolete. It has its place with new programmable controllers and control computers.

6

Central Air Conditioning Systems

John Haarhaus

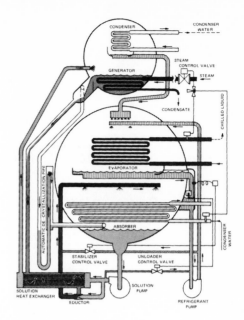

SCOPE OF CHAPTER

The central air conditioning system provides a stimulating challenge to the maintenance professional. The equipment represents a substantial capital investment. Subtle and seemingly insignificant symptoms may provide the only warning of disastrous failure. Minor repairs, when incorrectly performed, can trigger insidious attacks by invisible contaminants that may enter the internal system and go unnoticed until after the "autopsy."

Durability (expected service life) of central station equipment can be achieved only by preventive means, because irreversible damage can occur before any symptoms become obvious.

Durability, reliability, safety and efficiency are the ever-present goals of preventive maintenance. The relative importance you assign to each of these factors will dictate the character of your maintenance program. Sound judgment, based on knowledge and experience, will ensure its success.

The following sections on refrigeration compressors and absorbers provide a base of knowledge upon which you may structure your plan.

SECTION A
Absorption Air Conditioning

Jerome Morreale

One of the most widely used types of water chillers today is the absorption machine. It started to become most predominant in the air conditioning industry in the mid-1950s because of several factors:

1. Central air conditioning of large and high-rise buildings came into its own and absorption machines could be built that were capable of producing up to 1,500 tons of cooling capacity from a single unit.

2. The prime energy source for an absorption machine was low pressure steam (12 psi or less), although high temperature water could also be used.

3. Absorption machines run quietly and are vibration-free, which allowed them to be used for rooftop machine room installations in close proximity to cooling towers, thus reducing the cost of condenser water piping installation.

4. Absorption machines could be operated from power plant waste heat of exhaust steam.

5. Most large cities did not require a licensed operator for high tonnage absorption machines, thus eliminating the need for a skilled technician.

6. Large city utilities, such as Consolidated Edison in New York, promoted the use of absorption machines so they could sell central power plant steam during the summer months for cooling, as they did for heating in the winter. This also eliminated the need for boilers within the building and an operator to run the boilers.

Today there are thousands of absorption machines in use throughout the air conditioning industry. The major manufacturers of this equipment are the Carrier, Trane, and York Corporations. The operating cycles of all three manufacturers are basically the same, although machine construction and component design do vary. Carrier and York house the cycle components in a two-shell machine, whereas Trane employs the use of a single shell construction. (See Figures 6-1 and 6-2.)

BASIC CYCLE

The refrigerant in an absorption machine is water. This low-cost refrigerant is desirable for the absorption cycle because it can absorb energy by being made to boil at low temperatures, when subjected to an atmosphere that is maintained at a low

184

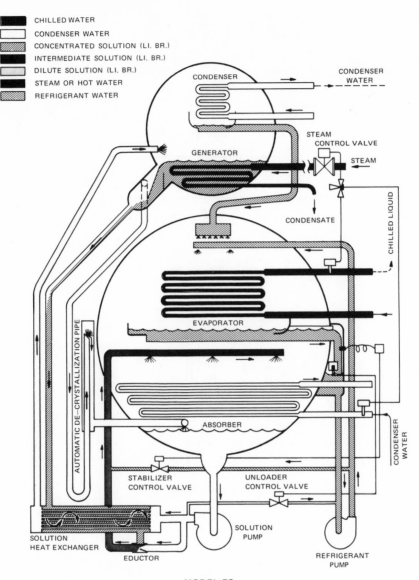

CHILLED WATER
CONDENSER WATER
CONCENTRATED SOLUTION (LI. BR.)
INTERMEDIATE SOLUTION (LI. BR.)
DILUTE SOLUTION (LI. BR.)
STEAM OR HOT WATER
REFRIGERANT WATER

MODEL ES
STANDARD STEAM CYCLE DIAGRAM

Courtesy of York Division of Borg-Warner Corp.

Figure 6-1.

York Absorption Cycle.

pressure or a high vacuum. For example; boiling point is 40°F at a pressure of .25 inches of mercury absolute. The absolute pressure scale is best described by referring to Figure 6-3.

The low pressure necessary for the refrigerant is maintained through the use

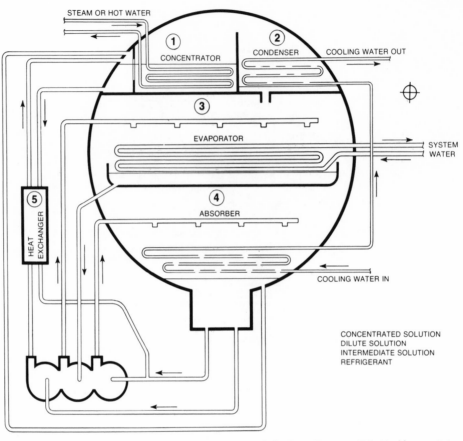

STEAM OR HOT WATER

① CONCENTRATOR

② CONDENSER

COOLING WATER OUT

③

EVAPORATOR

SYSTEM WATER

⑤ HEAT EXCHANGER

④ ABSORBER

COOLING WATER IN

CONCENTRATED SOLUTION
DILUTE SOLUTION
INTERMEDIATE SOLUTION
REFRIGERANT

©The Trane Company 1969. Used by permission.

Figure 6-2.

Trane Absorption Cycle.

of an absorbant, which in all cases is lithium bromide. Lithium bromide is a chemical that is mined and, after refining, is used in an absorption machine in a crystalline form dissolved in water. The lithium bromide solution has a great affinity for absorbing water vapor, thus making it a good absorbent for the cycle.

The absorption machine cycle is based upon two principles that bear repeating:

1. Water will boil at a low temperature when subjected to a high vacuum.

2. Lithium bromide solution has a great affinity for absorbing water vapor.

In Figure 6-4, we see a schematic depiction of the absorber cycle. Beginning with the evaporator section, the refrigerant (water) is subjected to a high vacuum, initially obtained by the use of a vacuum pump. The refrigerant is pumped through a spray header over the heat source, in this case the air conditioning load of the

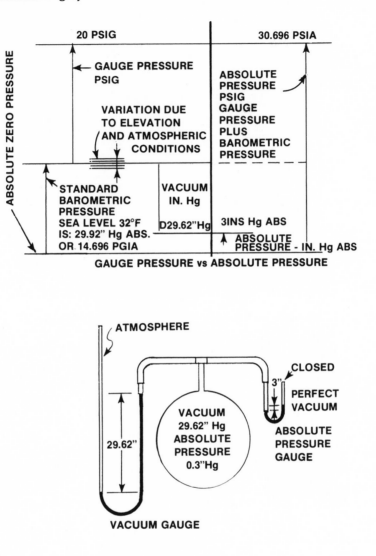

Figure 6-3.

Absolute Pressure Scale.

chilled water tube bundle. From there it absorbs heat energy, causing the refrigerant to boil and form water vapor, while chilling the water in the chilled water tubes. The water vapor must be removed immediately from this area to ensure that the high vacuum and low boiling temperature for the refrigerant can be maintained. Next, the absorbent, lithium bromide, which is housed in the absorber section, as seen in the diagram, is used. The water vapor in the evaporator section now flows to the lower pressure absorber area where it is absorbed by the lithium bromide, which is being sprayed through a spray header.

As the lithium bromide absorbs water vapor, it becomes more diluted, thus

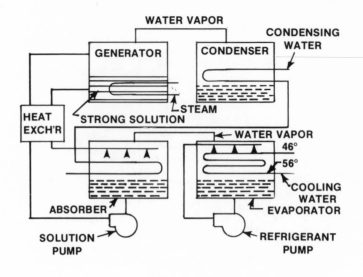

Figure 6-4.

Schematic of Typical Lithium Bromide Solution Cycle.

reducing its capacity to absorb. It is therefore necessary to reconcentrate the lithium bromide by removing the water. For this purpose, the third section, called the generator or concentrator is utilized. The weak lithium bromide is pumped through a shell and tube heat exchanger (to increase heat efficiency) to the generator section, where it is subjected to a heat source (steam or hot water tube bundle). This provides the heat energy to boil the excess water out of the lithium bromide, thus concentrating the solution. The excess water flows in the form of vapor to the fourth section, which is the condenser. This is provided with a cooling medium that, in most cases, is condensing water from a cooling tower. The water vapor is condensed and then flows back to the evaporator section, thus completing the cycle. The concentrated lithium bromide returns to the absorber section of the machine via the heat exchanger and is further cooled in the absorber section by the use of the cooling tower water, which is pumped through the absorber tubes prior to its use in the condenser. The cooling tower water flowing through the absorber tubes also removes the heat of the condensation, which is heat created when the lithium bromide solution absorbs the water vapor.

The four basic components of the cycle, the evaporator, absorber, generator (or concentrator), and the condenser, can be seen as they are actually housed in Figures 6-1 and 6-2.

CYCLE CONTROL

Today it is universally agreed that the best method of controlling the absorption cycle is to use a modulating steam inlet valve which is controlled by a thermostat, whose temperature sensing element is located in the chilled water line leaving the machine. Older methods of control, such as modulating condenser water flow or solution flow to the generator, have given way to the use of an automatic steam valve.

From the description of the absorption cycle, you can see how heat energy is transported from the inner spaces of a contained area, such as a building, via the closed chilled water system, through the absorption cycle, to the condensing water system, and on to a cooling tower, where it is rejected to the atmosphere. The absorption machine, which is a series of heat exchangers, is used as an intermediate transfer device. Certain essential functions became obvious from understanding the absorption cycle:

1. Services to the machine, such as adequate steam volume, proper condensing, and chilled water flow, are necessary for proper operation.

2. Heat transfer surfaces, both on the water and lithium bromide sides of the cycle, must be kept clean.

3. The machine must be kept airtight. Noncondensables that gather at the lowest pressure area of the cycle at the absorber cause an increase in vapor pressure, thus raising the boiling point of the refrigerant and subsequently lowering the capacity of the machine.

ESSENTIAL MAINTENANCE

As with any heat exchanger, the heat transfer surfaces must be kept clean. This begins with a proper chemical water treatment program for both the closed chilled water system and the open condensing water system. Strainers must be employed in both lines to protect these bundles, and these strainers must be kept clean in order to ensure adequate water flow. The condenser and absorber tubes should be cleaned with nylon brushes prior to each operating season to remove loose mud and dirt. At the same time, tube bundles and tube sheets can be inspected for hard scaling and signs of corrosion or erosion. The closed chilled water system can be cleaned less frequently. Since the absorption machine condenser tubes are made of copper, and absorber tubes are made of cupro-nickel, acid cleaning of these tubes should be avoided unless absolutely necessary and then done only by qualified personnel.

Nondestructive or eddy-current electronic tube testing is becoming a very common procedure for checking the physical condition and life expectancy of all types of machine tubes. These tests can measure tube wall thickness, indicating pitting, cracking, or tube wear, thus showing up potential problem areas and perhaps avoiding costly emergency tube replacements. Machines that are more than five years old should be tested every two to three years with accurate records maintained for comparison. Remember the tests should be performed only by trained, experienced personnel using the proper equipment.

No area of maintenance is more essential for proper absorption machine operation of the noncorrosive extended life of the lithium bromide circuit than the assurance of machine tightness. It is imperative that air be kept out of the vacuum side of the machine to avoid lithium bromide overconcentration during operation and to avoid corrosion at all times. During machine off-season shutdown periods, any areas of possible wear such as valve diaphragms, bellows assemblies, seals, etc., should be replaced in accordance with the manufacturer's recommendations. Only dry nitrogen should be used to break the vacuum in the unit during repairs. Pressure leak testing should be done after repairs, using an electronic leak detector and Freon-12 as a tracer gas to test all gasket joints, sight glasses, welds, etc. The machine should then be evacuated and subsequently able to maintain a standing vacuum during a prolonged shutdown period. During operation, the need for frequent purging is one indication of a possible air leak and immediate steps should be taken to effect repairs. Lithium bromide in the presence of oxygen is highly corrosive even if air is being purged through the machine in a small enough quantity not to affect cycle operation. The byproducts of corrosion clog internal passages such as spray nozzles, heat exchanger tubes and baffles, enough to decrease machine efficiency sometimes to the point of impairing total operation. Shell surfaces and tubes can become so adversely affected so as to destroy the entire machine long before its normal wearing life.

One simple test that can be used as a general indicator of machine condition, aside from the physical taking of cycle temperatures and pressures as recommended by the manufacturer, is a chemical lithium bromide analysis. In addition to measuring the corrosion byproducts washed in the bromide, the test measures the level of corrosion inhibitor still remaining within the solution, which can then be treated for maximum protection. The machine manufacturer or a qualified independent testing laboratory should be contacted to perform this test at least once per operating season.

All manufacturers that produce machines today use some type of hermetic circulating pump for lithium bromide and refrigerant water. Most of these pumps have replaceable internal bearings that should periodically be changed. Again, this should be done in accordance with the manufacturer's recommendations in order to protect against a mechanical failure. Electric motors should be megger-tested at least once a year with all line protection devices being regularly checked in order to avoid motor burnouts. Most of these motors cannot be rewound quickly in the field and can be replaced only at great cost and only where equipment is available.

While all absorption machines are similar in their operating cycle, their construction and hardware do vary. Some machines have open-type pumps that require seal changing, while newer units have the hermetic design pump previously discussed. Some manufacturers, such as Trane and York, use vacuum pumps for purge units, while Carrier has gone from the open-type of lithium bromide purge to the new hermetic motorless purge. Because of this diversification of hardware and design, general recommendations for shutdown overhaul and repair work are summarized below.

Vacuum Side of Machine

1. Take manometer reading of machine vacuum. On some units, an external manometer must be provided as it does not come with the machine. By

measuring the vacuum in inches of mercury absolute as compared to the saturation temperature for water at that pressure, you can establish a reading that can be used with a later reading, thereby giving a standing vacuum condition.

As shown in the chart below, a machine in a shutdown condition should have a reading of .8" HgA at an ambient temperature of 73°F. If the machine is allowed to stand at the same ambient temperature for a long period of time—a few weeks or a month—the same pressure reading should exist. If the pressure reading rises to 1.5" or 2" HgA at 73°F, you can assume that the machine has a leak. The pressure should rise only to a corresponding increase in ambient temperature.

VAPOR PRESSURE IN INCHES OF MERCURY ABSOLUTE	SATURATION TEMPERATURE
.18	32
.2	35
.25	40
.3	45
.4	53
.6	64
.8	73
1.0	80
1.5	92
2.0	102
2.5	109
3.0	115
3.5	120
4.0	125
5.0	134
6.0	141
8.0	152

This is a standing vacuum test and one of the easiest ways to check machine tightness.

2. Vacuum-test any control valve bellows assemblies. This test, using a standing water pipe or a vacuum gauge, should be performed with the valve energized and operating.

3. Inspect and/or replace all service valve diaphragms.

4. Inspect and/or replace all service sight glasses.

5. On older machines with removable spray headers for solution circuits, remove and clean these spray headers every two years. Newer machines with hermetic designs have welded-in spray headers that are not removed for normal maintenance.

6. Inspect all safety controls.

7. Check pump starters, clean contactors, overload relays, and magnetic starters.

8. Clean magnetic strainers in machine pump motor cooling coils on Trane machines where applicable.

9. Check pulldown efficiency of purge units in machines where vacuum pumps are used.

10. Clean and flush purge vacuum pump and charge with fresh oil.

11. Megger-test all motors.

12. Clean water side of condenser tubes every year.

13. Clean water side of absorber tubes every year.

14. Clean water side of evaporator tubes every three years.

15. Change hermetic pump bearings where applicable every two years.

16. Change seals on open pumps where applicable every two years.

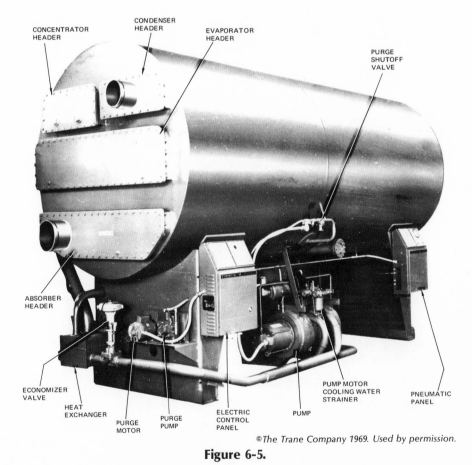

©The Trane Company 1969. Used by permission.

Figure 6-5.

Typical Single-Shell Absorption Machine.

Today, the most important consideration in the operation of absorption machines is energy conservation. Older machines, built in the late 1950s, operated at a steam rate of 18 pounds per hour per ton of refrigeration. Improvements in design have put today's machines at an operating level of closer to 10 to 12 pounds per hour per ton of refrigeration. The Trane Corporation now produces a two-stage concentrator, high-pressure machine with a 30 to 40 percent energy consumption improvement over the older single-stage machines. In addition, older Carrier, Trane, and York machines can be field-converted to operate with steam control and low temperature condensing water, substantially increasing machine efficiency and reducing steam consumption. However, the cost of such a conversion should be weighed against a reasonable economic payback as well as the overall machine condition and age. The maintenance work discussed in this chapter should be given every consideration, as it also contributes to economic energy conservation operation prior to higher cost conversion expenditures.

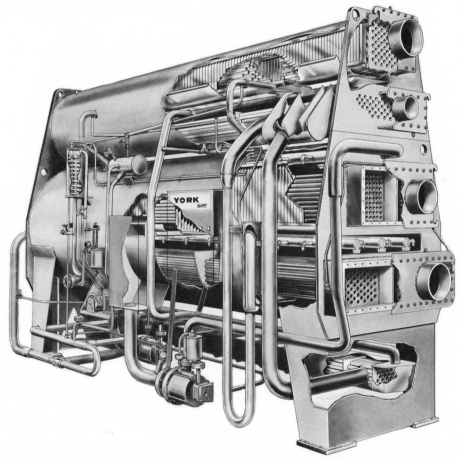

Courtesy of York Division of Borg-Warner Corp.

Figure 6-6.

Typical Two-Shell Absorption Machine.

SECTION B
Centrifugal Air Conditioning

Frank L. Phillips P.E.

GENERAL

There are two types of centrifugal compressors:

1. Open drive
2. Hermetically sealed

OPEN DRIVE CENTRIFUGAL COMPRESSORS

These compressors are driven by steam turbines or variable speed wound rotor electric motors through self-aligning couplings. Steam turbine drive is usually a direct drive requiring no increasing gear, since turbine speed matches the optimum speed of the centrifugal compressor and can be controlled.

Electric motor drive usually requires a speed increasing gear assembly since the synchronous speed of a four-pole or a six-pole electric motor—1,800 rpm and 1,200 rpm respectively—are too low for the centrifugal compressor and the full load speeds of these motors would be even lower, 1,725 rpm and 1,125 rpm.

Electric motor drives are usually wound rotor machines controlled by resistance inserted in the rotor circuit through a drum-type controller to allow efficient, variable speed control of the centrifugal compressor to suit load conditions. Drum control can be either manual or automatic.

Capacity Control

Controlling compressor speed is the most efficient method of capacity control for power conservation:

1. by means of a speed controller and turbine governor to vary turbine speed.
2. by variable resistance drum controller for wound rotor induction motor to vary rotor speed.

194

3. by magnetic or hydraulic coupling where synchronous and squirrel cage constant speed electric motors are used to vary compressor speed.

4. by throttling compressor suction by means of a thermostatically controlled suction damper operating in response to chilled water temperature. Variable inlet guide vanes are sometimes used in the first stage position of the compressor for control below 10 percent capacity. Capacity of a centrifugal machine is controlled to maintain the correct water or brine temperature to suit load conditions. Therefore, the best control from the standpoint of power input is the speed control method.

These methods of controlling compressor output are effective from 30 percent of full capacity to full load. Below 30 percent capacity a hot gas bypass is usually provided, except on machines using the variable inlet guide vane control.

HERMETICALLY SEALED COMPRESSORS

These compressors are similar in most respects to the open drive centrifugal compressors, except they are limited to electric drive and motor; speed increasing gears and the compressor are all hermetically sealed in an enclosure with proper lubricant.

FUNCTION OF COMPONENT PARTS (See Figure 6-7)

Chiller

Heat exchanger that cools water in tubes by evaporation of liquid refrigerant in shell surrounding the tubes.

Compressors

Centrifugal pump that reduces cooler pressure to facilitate evaporation of refrigerant and compresses gas and delivers it to the condenser at high pressure and high temperature.

Condenser

Heat exchanger that cools the high temperature gas from the compressor to condense it and causes it to change to a liquid state. Condenser cooling water from cooling tower or other source passes through tubes and refrigerant remains in the shell of the condenser surrounding the tubes.

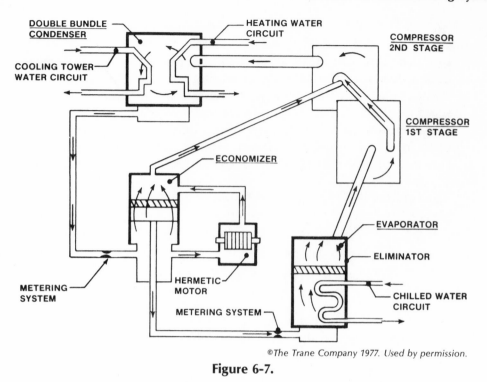

Figure 6-7.

Diagram of Typical Centrifugal Machine with Compressor Chiller-Condenser.

Controls

Automatic devices for starting, stopping, and controlling the speed of the compressor and auxiliary pumps, fans, and equipment, and for operation and protection of component parts of the machine such as water temperature, pressure, and flow.

Drive

Electric motor or steam turbine connected through a universal coupling to the compressor drive shaft or drive shaft of the speed-increasing gear unit.

Economizer

Device to further cool liquid refrigerant from condenser by partial evaporation. The cooled liquid refrigerant returns to the cooler and the evaporated portion passes back to the compressor.

Purge Unit

Scavenger unit condensing separator that removes air, water, and noncondensables from the refrigerant. This unit automatically vents air and gases, and collects water for removal by the operator.

TYPES OF PURGE UNITS

Thermal

The thermal purge operates automatically to remove water, air and noncondensable gases. Shutoff valves in the purge unit should be open at all times. Operating and safety switches and devices are usually factory set but should be checked periodically with the manufacturer's data that is furnished with the machine.

Motor Driven Compressor-Type Purge

This system is usually furnished with machines designed for low temperature duty and should be operated continuously whenever the refrigeration machine is in operation. All valves on this type of purge unit should be open at all times with the exception of the water and oil drain valves.

Purge pressure and temperature settings are adjusted at the factory but should periodically be checked and reset in accordance with the manufacturer's data, which is usually furnished in table form with the machine.

Oil level is visible in the compressor sight glass both during purge operation and shutdown. Oil level should be maintained by adding oil according to manufacturer's specifications furnished with the unit. Before adding oil, the compressor must be shut down and the valve in the liquid refrigerant return line *closed*. Oil can then be added, usually by removing a plug at the top of the compressor sight glass.

Motor-driven purge units can be used to evacuate or pressurize the machine if the machine is not equipped with a pump-out unit.

PURGE SYSTEM

The purpose of the purge system is to keep the system clean and free of excess air and water. Air in the machine will raise condenser pressure, which, in turn, will raise the compressor load and horsepower input. High condensing pressure increases energy costs and lowers capacity.

Therefore, the purge unit protects the component parts of the machine, since it operates as a good air and moisture scavenger and a good indicator of machine tightness.

Water present even in small quantity when mixed with refrigerant can cause hydrochloric acid and hydrofluoric acid to form. This attacks and corrodes the metal parts of the machine and causes sludge to form in the lubricating oil.

MAINTENANCE OF PURGE UNIT

- Periodically clean all strainers.
- Periodically clean float valve or metering device.
- Periodically check and adjust controls.
- Keep sight glass clean—if presence of water is indicated, drain the water, measure the quantity drained, and record the amount for future reference.

CONDENSER MAINTENANCE

- Periodically check the purge condensing chamber and clean the chamber.
- Periodically check the oil level of a motor driven purge unit.
- Maintain a tight condenser and clean condenser tubes annually.
- Provide water treatment if the condenser water is recirculated.

OPERATION

Both types of purge units return refrigerant to the cooler or an area in the economizer where temperatures are above freezing. If both these areas are below freezing, the back pressure regulating valve in the purge return line will operate to prevent the purge unit from freezing.

This back pressure valve should be adjusted to the manufacturer's specifications furnished with the unit.

Should condensation of refrigerant occur in the suction line between the purge compressor and the condenser with compressor-type purge units, the condensate may enter the crankcase, causing oil to "foam" and resulting in loss of oil. The pressure-reducing valve should be readjusted in accordance with the manufacturer's recommended settings furnished with the unit.

DAILY LOGS

The most important part of any operation and maintenance schedule is the method of recording data. A daily log should be started with the start-up of the machine and should record:

- Whether or not there is a purge discharge immediately after start-up that could indicate leaks during shutdown period.
- Frequency of air discharged by the purge unit—once or twice a day is normal. More frequent discharges while machine is in operation indicate leaks.

- Amount of air discharged can be observed by directing purge discharge into a container of water. (The manufacturer's manual will list several methods of testing for air and water.)
- Water in the purge separation chamber indicates refrigerant has reached saturation. Water must be removed manually by the operator and this operation makes the operator aware that water is present.
- The operator must then take the necessary steps to determine the source of water leaks and effect repairs.

Purge units should be operated during long periods of shutdown to determine if there are air leaks and/or water leaks.

The operator should be aware that the automatic exhausting of air and noncondensables by the purge unit also causes some refrigerant loss; some refrigerant gas is always present, since condensing of the refrigerant is not always complete. The operator should therefore check the refrigerant level from time to time.

DAILY INSPECTION AND MAINTENANCE

Make sure that the machine is clean and tight. The presence of foreign material or evidence of leaks should be investigated immediately to determine the source and cause, and remedial action should be taken to effect a clean, tight machine.

- Check for proper purge operation as a further check on machine tightness.
- All safety controls should be checked and kept in first-class working order.
- Check lubrication and be sure to maintain a clean lubrication system.
- Keep adequate operating records for a day-to-day comparison that will indicate changes in the efficient operation of the equipment.

PERIODIC OR ANNUAL MAINTENANCE

- Clean heat exchange tubes.
- Check refrigerant level.

Since some refrigerant is lost when relief pressure devices operate to relieve pressure, set the relief pressure devices high enough to minimize refrigerant loss.

LUBRICATION

Bearings last longer when a good, clean lubricating system is maintained. The lubrication system is affected by heat, dilution (with refrigerant), and moisture. The operator should keep the lubrication system clean at all times and use only the lubricant specified by the manufacturer. This is usually a good quality turbine oil especially treated and fortified with inhibitors to minimize oxidation and foaming according to the type of compressor, evaporator, temperature, etc. Oils should never be mixed, or oils used with a given machine used in another machine unless their characteristics are similar in all respects.

Change oils once a year or every time the centrifugal is opened for inspection or repair. Seals on the oil container should not be broken until the operator is ready to refill the oil container with new oil.

Have oil sample analyzed after each oil change by an experienced laboratory to detect any unusual conditions which might indicate excessive wear, contamination, improper lubricant, or defects in the lubrication system.

OIL PRESSURE

Never maintain oil pressure higher than the manufacturer's specifications. Oil pressure should be noted in the daily log and the oil-regulating valve adjusted if necessary to maintain correct oil pressure.

OIL TEMPERATURE

Correct oil temperature should be maintained, during both machine operation and shutdown, to minimize the amount of refrigerant absorbed by the lubricating oil. If the lubricating oil heater or the controlling thermostat does not function correctly as per the manufacturer's instructions, a high concentration of refrigerant can occur during shutdown, which will cause a very high release of refrigerant at start-up. The result is oil foaming, decreased oil pressure, loss of oil, and unnecessary "trip outs." Such conditions, including the frequent restarts, cause excessive bearing wear and deterioration.

OIL COOLERS

Oil coolers are usually furnished with centrifugal machines but, if not, they can be added. Such devices consist of a shell and tube heat exchanger utilizing a small quantity of cooling water to maintain normal lubricating oil temperatures.

OIL FILTERS

Oil filters should be replaced whenever lubricating oil is changed. Daily logs taken at start-up and at least once during the daily operating cycle will record machine operation, tightness, changes in power input, purge operation, refrigerant loss, oil temperature, and other pertinent data, which will indicate to the operator the general operating condition of the equipment and aid him in analyzing machine performance. Conditions that may cause trouble can thus be corrected before a major breakdown occurs. This method serves as an excellent guide for planning preventive maintenance.

CONCLUSION

For the average operator, a basic knowledge of the component parts of a centrifugal machine and the functions of each component part are the first requisites for good maintenance and economical operation of the equipment. If the operator possesses this knowledge and applies his own inherent common sense, he will be well on the way to operating an efficient, trouble-free plant.

The operator should confine himself and his work to operating and maintaining a clean, tight, well-lubricated machine, and the keeping of accurate logs to monitor machine operation. Minor adjustments and repairs indicated by the logs can be carried out by the operator at convenient times using the manufacturer's manual as a guide. However, extensive repairs should not be attempted by the plant operator, only by a qualified technician. Most major manufacturers offer a service contract that provides one or two visits to the plant per year and covers the cost of normal adjustments and minor repairs, including parts and labor.

SECTION C
Reciprocating Air Conditioning

Robert M. Barr

SCOPE OF CHAPTER

Contrast these scenes:

- Year—1950.

- Place—The equipment room in a sizable building on a university campus.

- Fifty tons of air conditioning are supplied with a built-up system, using R12 refrigerant. The compressor is an open type, belt driven with a 65-horsepower motor at 1,150 rpm. The motor compressor is mounted on a stand in the center of the room, accessible from all sides. Any breakdown can probably be repaired in place, since it will affect only the compressor and contaminants are not likely to enter the refrigeration cycle.

- Year—1980.

- Place—The same.

- Fifty tons of air conditioning are supplied with two package systems, each with a 25-horsepower semi-hermetic compressor turning at 1,750 rpm, using R22 refrigerant. Both units have been squeezed into a corner to make room for other equipment acquired over the years. A breakdown now will very likely mean a motor burnout, and contamination of the entire system is almost certain. There is no room to work on the compressor, even if the time and skilled manpower were available. A replacement compressor will be required, along with necessary filter driers and oil changes to remove the contamination.

These two scenes portray the basic changes that have taken place in the design approach of reciprocating air conditioning systems over a 30-year period. The bottom line is that the semi-hermetic compressor is literally twice as likely to fail as its open predecessor. Insurance premiums verify the fact that the cost of failure is twice as high on semi-hermetic machines. But a word of caution is necessary. The term *fail* implies some deficiency on the part of the compressor itself, as if its design were inadequate. Such an implication is unfair and untrue. The majority of compressors manufactured today have a realistic life span of seven to eight years, when used in air conditioning applications in the northern United States. In heavier usage, such as in Florida, only a year or so would need to be deducted, assuming the same use. In an application such as a computer room, where the machine operates virtually continuously, the actual number of anticipatable hours would rise, though the number of years would decline.

Only in a very few cases does the compressor simply *fail*. Much more frequently, it has been subject to some abuse or lack of care or misapplication, which causes failure. The fault, almost always, is in the *system* rather than in the compressor. This chapter will deal with such abuses and how to avoid them.

PROTECTING THE COMPRESSOR FROM ABUSE

There are two stages that require protection. The first is in the original installation. The following is a simple checklist to help sidestep major problems. Of course, the manufacturer's instructions for a particular unit should be followed in detail.

1. *Sizing*

   Avoid continuous operation of a refrigeration compressor at minimum load. Oversizing leads to a variety of problems in the system and with the compressor itself. Among the more obvious are: poor oil movement at low load, (low refrigerant velocity leading to poor oil return from other parts of the system); coil icing, (due to suction temperature dropping too low); frozen coolers, (for similar reasons); and compressor overheating (low gas volume at high suction temperatures often causes motor failure in semi-hermetics despite the presence of thermal overloads). Most manufacturers provide some form of capacity control accessory to accommodate load variation, and one of these should be employed in the original system design.

2. *Vibration and Mounting*

   The manufacturer's recommendations should be followed carefully, since vibration causes annoying noise and can also lead to serious mechanical problems.

3. *Space and Ventilation*

   Sufficient space should be available all around the compressor to permit work to be done without undue restriction. The maintenance manager should be consulted at the time of the original design; and the possibility that load and/or equipment may be added as time goes by should be taken into consideration.

   In the case of water-cooled systems, care should be taken to provide enough end space for cleaning and replacing condenser tubes.

Ventilation is often mandated by local codes, but imagination and common sense will often suggest ways of relieving problems. For example, building exhaust air might be passed through the compressor area; or, conversely, the equipment room might be used as the air intake plenum. Forced ventilation may be used where natural ventilation is impractical. In any case, excessive temperatures will shorten the operating life of open compressor motors. Since semi-hermetic compressor motors are cooled by suction gas, they pose fewer problems in this area.

4. *Alignment*

Alignment is a problem peculiar to open compressors, whether direct coupled or belt driven. It is critically important that the installer align coupled machines with a dial indicator. It simply cannot be done by visual inspection. Misalignment can result in rapid wear and deterioration.

5. *Exposure to Atmosphere*

The compressor leaves the factory sealed against moisture or other atmospheric contamination. Care should be taken not to remove these seals until piping is to be connected. Later discussion will clarify the importance of eliminating moisture.

6. *Interlocks*

The control system must be set up so that the auxiliary contacts of the system fan (or of the chilled water pump starter) on the one hand, and the condenser water pump starter on the other hand, are *closed* before the compressor can be started. Such an interlocking circuit will provide that, should either starter fail to energize or become de-energized by any of its safety controls, its auxiliary contacts will open, thus stopping the compressor. The same kind of interlock should be provided for the exhaust system—so that (typical of a hospital) in the event of a fire alarm, which stops all exhaust activity, the compressor is also de-energized.

7. *Separate Wiring for Crankcase Heaters*

A crankcase heater is designed to minimize the collection of refrigerant in the crankcase by warming the oil enough so that refrigerant boils off. But it is, typically, only about 100 watts—similar to an average light bulb. It requires, therefore, a fairly long time to heat up the reservoir of oil; it is simply not possible to turn it on and to expect that the crankcase will be freed of refrigerant quickly. The safest course is to permit the crankcase heater to stay on all the time, except when the compressor is actually running. The electric consumption is quite small and the additional margin of safety quite large. It should be confirmed that the crankcase heater is wired independently of the on-off switch used to shut off the compressor for the winter.

8. *The Air Filters*

They are generally of the throwaway type and quite inexpensive. If there is the slightest question of their being dirty from construction activity, change them to assure good air flow through the direct expansion air cooling coil.

9. *Suction Line Drier*

    A temporary suction line drier (replaceable core type) in the system at the original installation is good extra insurance against compressor damage. Check the pressure drop, however, to be sure it does not exceed the design. Cores should be removed after the system has been proved out.

10. *Wiring of Controls*

    This would apply especially to such items as oil pressure switches and flow controls on chilled water systems. A few minutes to check that they are properly wired may save much time and expense later.

START-UP AND OPERATION

The second stage has to do with start-up and operation, not only the initial start, but with each seasonal start as well.

1. *Air Handling Equipment*

    Are all belts in place and in good shape? Are they aligned and properly tightened? Are bearings lubricated? Are air passages clear—unblocked by cartons and trash? Have filters been checked and replaced if necessary?

2. *On Water-Cooled Systems*

    Has the sump of the tower or evaporator condenser been cleaned, leak checked, and filled with water? Has water treatment been initiated?

3. *Valves*

    Are they tagged and have they been properly opened? This applies not only to the shut-off valves of a water-cooled condenser, but to the service valves of a compressor, both suction and discharge. Operating the compressor with valves closed will result in rapid overheating and seizure.

4. *Controls*

    Are they set to automatic, where this is required? This applies to liquid line solenoids and interlock circuits.

5. *Electrical*

    Have controls been visually inspected for pitting or welding on starters and/or for obviously worn components? Is the crankcase heater working? Has it been on long enough to heat the oil and drive out the refrigerant?

6. *After Start-up*

    Once the system has operated for 15 to 30 minutes, it may be assumed to be in equilibrium. The compressor oil level should then be checked at the sight glass; the oil pressure checked with a gauge; and the liquid line sight glass checked visually.

It is obvious that, once the system is on line, periodic checks must be taken. But how frequently? Given the trade-off between cost and safety, it is suggested that a thorough monthly check is better than more cursory weekly checks. To that end, the following checklist should be followed.

PREVENTIVE MAINTENANCE

This monthly checklist *must be logged* and kept as a permanent record. It is simple, its cost is negligible and its potential benefits are large.

1. Read and record refrigerant gauge pressures.

2. Read and record oil pressure gauge reading.

3. Check compressor oil level at sight glass and record.

4. Take sample of compressor oil and conduct a "one time" acid test. Record result. (This costs about $3.00, but it is the best early warning signal of moisture.)

5. Check and record refrigerant charge at liquid line sight glass.

6. Check and record suction gas temperature for proper superheat. (It is suggested that superheat setting be locked down and painted over with red nail polish. If it is ever necessary to change it, it will also be necessary to replace the expansion valve.)
    See section on page 207 for Setting Superheat

7. Check crankcase heater and all pressure switches to be sure they are working. Record.

8. Check air filters and change if necessary. Record.

9. Check and tighten or lubricate, if necessary, bearings, belts, fans. Record.

10. If you find a leak of any kind—refrigerant, water, steam—stop and repair it. Record. Leaks are costly and may be dangerous.

PURPOSE OF THE COMPRESSOR

The purpose of the refrigeration compressor is to pump and compress refrigerant gas in order to raise the pressure of that gas from the evaporator pressure to the condensing pressure. When it functions properly, it delivers refrigerant to the condenser at the correct pressure and temperature for the condensing process.

COMPRESSOR FAILURE: THE PRINCIPAL CAUSES

Liquid Slugging or "Floodback"

In the refrigeration cycle, it is essential for the flow of refrigerant to the evaporator to be controlled. Too little will provide inadequate cooling; if there is too much, some will pass through unevaporated and enter the compressor suction line. The compressor is designed to pump *gas,* not refrigerant in liquid form. Liquid, being noncompressible, can easily destroy the compressor. The metal parts will break upon impact with a noncompressible liquid.

Oddly, while valves often break first, there are instances when valves remain intact while pistons, rods, and rings break up. In any case, this type of failure is often accompanied by considerable noise as bits and pieces fly around inside the compressor; in extreme cases, the casing itself breaks or a crankshaft snaps, or (in semi-hermetics) the motor burns out because of a stray piece of metal lodging in the windings.

The most common cause of liquid slugging is improper setting of, or malfunction of, either the superheat or the thermostatic expansion valve. (See Figure 6-8—The Principal Parts of a Thermostatic Expansion Valve and Figure 6-9—Diagram of an Evaporator Coil.)

The Expansion Valve

An expansion valve is a relatively simple mechanical valve. The bulb and capillary tube contain a gas or liquid, usually the same refrigerant used in the system. When heat is applied to the remote bulb, the pressure within increases and is transmitted to the area above a diaphragm. The diaphragm flexes downward against the tension of a spring, pushing a pin away from its seat and causing the valve to open. The adjustment changes the tension of the spring; this variability is referred to as the *superheat adjustment.*

> Note that the remote bulb location is important. Usually it is placed on the suction piping, near the outlet of the evaporator coil.

Setting Superheat

Superheat is the *difference* in temperature between the evaporating temperature and gas temperature at the remote bulb. This difference remains relatively constant regardless of the variations in the evaporating or suction temperatures.

If superheat is set too low, raw liquid can return to the machine with the damage as explained previously. If superheat is set too high, the motor windings of a

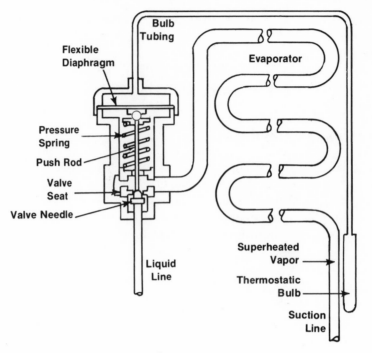

Figure 6-8.

The Principal Parts of a Thermostatic Expansion Valve.

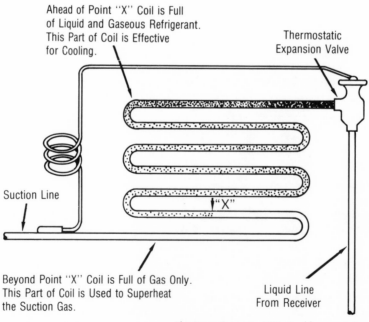

Ahead of Point ''X'' Coil is Full
of Liquid and Gaseous Refrigerant.
This Part of Coil is Effective
for Cooling.

Thermostatic
Expansion Valve

Suction Line

''X''

Beyond Point ''X'' Coil is Full of Gas Only.
This Part of Coil is Used to Superheat
the Suction Gas.

Liquid Line
From Receiver

The Trane Company 1977. Used by permission.

Figure 6-9.

Diagram of an Evaporator Coil.

hermetic compressor will not be cooled sufficiently by the returning gas and may burn out. In an open compressor, the overheating will generally occur in the oil, at points of lubrication, often causing seizing.

In Figure 6-9 you will see that Point X changes if the superheat setting is changed and, as noted, the part of the coil beyond Point X is useless for cooling purposes. For this reason, superheat must be set properly to obtain proper evaporator performance without exposing the compressor to liquid floodback.

A special note of caution is necessary here, since most air conditioning craftsmen have been taught to measure temperatures for this purpose *at the evaporator coil,* on the theory that to do so gets maximum efficiency from the coil. While that much is true, it tends to obscure the real purpose of superheat: *to protect the compressor.* Temperatures and pressures to set superheat should be taken *at the compressor first.*

The manufacturer's instructions should be followed exactly with respect to where to set superheat for his equipment. Most thermostatic expansion valves are adjustable over a superheat range of 0° to 25°-30°.

It is also necessary to recognize that—once it is properly set up—it should not be necessary to change a superheat setting. If something happens that requires a superheat adjustment, it is probable that something is wrong with the expansion valve itself. It may have sprung a leak in the bulb, it may have been plugged up with wax from the oil, or it may have hung up on copper plating as a result of some acid in the system.

> **A good rule: If you have to change the superheat setting, replace the expansion valve.**

Low Oil Pressure

A second major cause of failure is low, or no, oil pressure. Occasionally a compressor will leave the factory without oil, or an accident in transferring it from place to place drains it. Remember that some oil always moves around the system with the refrigerant. But if there was a correct oil level in the system originally, it should return to the compressor unless something is wrong, such as:

1. *Oil pump failure*
   This can be anything from a broken tang to the reversing gear being stuck. Sometimes just reversing the direction of compressor rotation by switching the motor leads will do the trick.

2. *Oil relief valve blocked open*
   A little piece of dirt can cause big trouble.

3. *Worn bearings, worn unloader assembly, broken oil tube*
   A number of internal problems of this kind can occur. In most cases, the most economical solution is to exchange it for a rebuilt compressor.

4. *Excessive refrigerant in the crankcase*

This can occur during layoff periods or in swing seasons because of migration. Migration is the tendency of a refrigerant to move to and condense in the coldest part of the system. Since *cold* is a relative term and the difference may be only a few degrees, this phenomenon often occurs after an overnight layoff. For example, the compressor, when operating, tends to run at a fairly high temperature; the heads will literally feel hot. Since it is made of cast iron, it is also quite heavy in relation to its surface. Therefore, by definition, it is a heat sink: it is the last part of the system to cool off at night, but once down to ambient temperature, it is equally slow to heat up. As the ambient rises (for example, on a sunny spring morning) the compressor warms along with the rest of the system, but lags behind due to its mass, and it can become the coolest part of the system just at the time when the thermostat is calling for cooling. For this reason, noise and vibration from the compressor during a morning at start-up can usually be attributed to refrigerant in the crankcase.

Check the crankcase heater for proper operation. Check the liquid line solenoid valve and be sure the system is not overcharged.

The basic rule is: refrigeration oils absorb refrigerants at all temperature ranges; this absorption must be controlled. Oil will move around the system, but *must* be able to return to the compressor.

BURNOUTS

This section refers to semi-hermetic compressor failures only. Open motors just do not burn out due to compressor failures. Even if an open compressor seizes up, the coupling is likely to break first, before the motor fails.

Principal Causes of Burnouts

1. Low Voltage

In recent years, this has become increasingly more likely because power companies have had to cut back on voltage in peak periods. When the voltage drops, the amperage goes up. Within about 10 percent leeway, the motor can probably take it; beyond that, failure is likely. Overloads built into the winding—whether simple temperature-operated switches or fancier solid state devices—are only a partial answer. The overload covers a relatively small part of the winding and unless the overheating takes place in that particular area, the motor may be burned before the sensor can function. Sometimes, just a bad contactor gives low voltage at the compressor itself, although the voltage being delivered by the power company is all right.

2. Failure to clean up a system from a previous burnout

This happens surprisingly often, considering that so much attention has been given to clean-ups. When there have been several burnouts in a row on the same unit, it seems obvious that the fault lies, not with the compressor, but with the system.

Not only must a suction line drier be put on (and, preferably, the cores changed several times) but the compressor oil should likewise be analyzed each time the drier core is changed. The oil should be changed when indicated by analysis or when its condition is doubtful.

Simply blowing out a system with Freon will not do. The carbon and acid associated with a burnout will find its way into every crack and crevice in the interior of the compressor body and into every bend and valve in the piping system. If not cleaned out thoroughly, the Freon (one of the world's great solvents) will eventually dissolve one of these clumps and carry the material back into the stator where it will again eat through the insulation and burn out the new windings.

Cleaning out acid and carbon can be done effectively with a liquid line drier and a suction line filter-drier but the clean-up should be confirmed by an oil analysis.

The concept of flushing out with R11 is outdated and is not recommended for the following reasons:

■ it does not get out all the contaminants, but simply shifts their location from one spot to another in the system,

■ it is difficult to remove all the R11 from the system, and

■ even small amounts of R11 left in the system may cause adverse chemical reactions.

By the same token, it must be recognized that just putting in a drier core and walking away is not enough. Oversized drier shells can and will trap oil. The oil level in the compressor should be checked carefully while the clean-up driers are in place. If a drop in oil level is noted, it tells the craftsman to remove the drier shells and correct the oil return problem.

Every manufacturer and rebuilder requires that a proper clean-up be performed in order to validate the warranty on an exchange compressor.

3. Spot Burnout

In this case, the damage is usually confined to a small area. It might happen as a result of a short between windings, or even between wires within a winding. It might be from windings to ground. Several causes are possible: faulty manufacture, mishandling (stator dropped, for example, which will leave a flattened spot), or deterioration of insulation caused by age. But, it might also come about as a result of a metal particle broken loose during a liquid slug and subsequently shorting the winding.

4. **Single-phase burn**

   This happens when two phases of a three-phase motor burn because they receive the full current flow. Most frequently, this results from malfunction of contactors or other electrical failure. It is characterized by a "striping" effect at the end of the stator.

5. **General Burnout**

   Often it is difficult to pinpoint the cause because literally nothing is left. Generally, a strong smell accompanies a burnout. Remember: many burnouts are caused by moisture in the system.

MOISTURE

Much has been written about moisture in an air conditioning or refrigeration system, but it all boils down to one thing: *Moisture is an absolute enemy and must be eliminated.*

The reason is simple. Moisture (even in very tiny amounts) combines with Freon to make formic acid. For example, in a typical refrigeration system, damage from moisture in R12 can occur from solubility as low as two ppm (two parts per million).

In short, all the normal precautions should be taken during installation to ensure that moisture does not enter the system. Normally, this presents no problem to the craftsman—the problems tend to arise later, when the system is in service. Even though moisture will not usually enter a system while it is pressurized, it can enter in the process of repairing a leak when it is necessary to pump down the compressor and remove the system gas. When the system is at atmospheric pressure, moisture is very likely to enter.

A Sure Sign of Moisture

When a compressor is opened and disassembled, there is never a question as to whether there has been moisture in the system. If moisture has been present for any significant period of time, at least some parts (often connecting rod inserts, valves or bearings) will be copper-plated—literally, they will look as if they were coated with copper. This electrochemical reaction is due to the presence of formic acid, which in turn is a clear sign that moisture was present. Should the gas side of the system prove to be tight and dry, it would be logical to look for leakage in a water-cooled condenser as the source of moisture.

Removing Moisture

A good vacuum pump is an absolute necessity. (Caution: do *not* attempt to use an air conditioning compressor to evacuate the system; it will not attain a satisfactory vacuum, and it may contaminate the system). A good vacuum pump should be capable of evacuating to an absolute pressure of not higher than 500 microns of mercury (one half millimeter Hg).

How much vacuum is necessary? Like the perfect recipe for a martini, opinions vary. The standard triple evacuation technique calls for dropping to 1,500 microns the first two times and to 500 microns the third, while breaking the vacuum to a positive pressure with the refrigerant after each evacuation. On large systems, it may be wise to get below 150 microns. A larger vacuum pump will help reduce the amount of time required, but the size of the connection between the pump and the system is even more important. Put simply, it should be as large a connection as possible in diameter, as short as possible in length, and connected to both the low side and the high side. The vacuum gauge (usually a compound gauge) should be put on the system as far away from the pump as possible.

AIR IN SYSTEM

Pulling a vacuum, of course, removes the air in which moisture is carried. But even if the air were perfectly dry, air itself is a contaminant that can elevate not only the condensing temperature, but can, through its oxygen content, create a chemical reaction with the oil and refrigerant. Usually, a sign of air in a system is the blackening of the discharge side of valve plates, since the discharge valve is the hottest point.

Leak Detection

Only a direct visual indicator (a soap solution with a little glycerine added to retard drying) or a sensitive instrument should be used to detect gas leaks in a refrigeration system. It is impossible to detect a leak by pressurizing a system and then attempting to read a pressure gauge.

Of the three common types of instruments available, the halide torch is least expensive, but suffers from being unusable in a windy or saturated atmosphere. An electronic detector is even more sensitive, but likewise cannot be used where considerable leakage has taken place. Most sensitive is the sonic type, and it has three other advantages to offset its cost—it can be used at a considerable distance from the actual leak, it will function in a saturated atmosphere, and it can be used in a vacuum.

Repairing Leaks

The presence of a leak leads to the suspicion that there may be more, so that every possible point should be tested. Moreover, every leak found must be repaired. In some cases this may require new fittings; in others, new flared or soldered joints; and in others, new gaskets or packings. Whatever is required must be done. *A Freon leak is intolerable: It not only wastes expensive gas but, if not corrected, will certainly lead to worse problems.* Remember that outward leaks may appear in Freon systems on both the high and low sides of the system, since both may be above atmospheric pressure.

Suggestions on Soldering

Fluxes containing acids should never be used to make R12 or R22 refrigeration joints; the acidic ingredients will contaminate the system and can damage the compressor severely.

Flux is applied only to the outside of the tubing on the male end of the fitting—the end is never dipped into the container of flux. Such fluxes, when heated, become hard and brittle as glass. If allowed on the inside of the joint, particles would break off and eventually find their way into the compressor.

Good soldering technique usually requires the use of an oxyacetylene torch and silver solder. When extensive soldering is to be performed, the passage of an inert gas, such as dry nitrogen through the tubing, will prevent the formation of damaging oxides on the inside of the joint.

Testing for Integrity

Once convinced that the system is leak free, the craftsman should conduct an overall test for integrity. He will pull a vacuum to below 250 microns and blank-off the system. Then he will close the valve and watch the pressure on the gauge. Either it will increase for 15 to 30 minutes and then stop (which means the pressure in the system has equalized and there are no leaks), or it will continue to increase (which means that leaks still exist and must be found and corrected). *Never charge a system with Freon while there is still a leak in it: That will only compound the problem.*

ELECTRICAL PROBLEMS

These generally break down into four categories:

1. Abnormal supply
2. Malfunction of some electrical component or accessory
3. "Jumping" a control
4. Improper wiring

Obviously, it is also possible to have a manufacturer- or rebuilder-related problem, but the vast majority are field-related.

Abnormal Electrical Supply

Generally, this is a voltage problem—either high or low. The motor winding temperature will tend to rise on either side of the normal line voltage, but low voltage tends to be the more prevalent cause of motor failure. As noted, the safe range allowed by manufacturers is +10 to −10 percent. It should be recalled, however, that when the demand exceeds available supply, the cutback made at the powerhouse may be within acceptable limits; but by the time it reaches the field location, the voltage may be significantly below minimum requirements because of line losses.

Other problems also exist, notably unbalanced phases that show up as variations in amperage and/or voltages of the three legs. This can be sufficient to damage a motor winding. Frequently, only two legs of a three-phase motor have had overload protection. This leaves a motor vulnerable to damage by single-phasing, due to failure of the unprotected phase. It is good practice to use phase failure protection *on all three legs* of the power supply.

A rule of thumb is to not exceed 10 percent variation in amperes from leg to leg; should it be higher, check the voltages on all three legs—they should not vary more than 2 percent. If more, contact the power company. However, if voltages are virtually equal, switch the leads on the highest and lowest amperage legs; if the high amperage stays at the same terminal, it is the motor that is at fault. On the other hand, if the high amperage moves with the lead, it is the power supply.

Malfunction of Electric Component

Visual inspection can often detect possible failures—for example, pitting, welding, or corrosion on starter contacts. If a starter doesn't look perfect, replace it—it's far cheaper than replacing a compressor. The same is true of relays; contrary to popular opinion, they do wear. A good rule of corrective maintenance is to replace starting relays whenever you suspect deficient performance. Another good rule is to replace the starting contactor any time you replace the compressor. Be sure that contactors, controls, fuses, and size of wires used are correct for the nameplate and the locked rotor amperage (LRA) of the motor. Chattering controls can produce high transient voltages, which may cause motor winding breakdown.

"Jumped" Controls

One of the more common failures in air conditioning is caused by bypassing a control that is *known* to be inoperative. The craftsman jumps a defective oil safety switch, with the intention of coming back the next day with a replacement control. But he is delayed and ultimately forgets. Subsequently, a condition arises that should have caused this control to shut down the system; but, it can't do so, and a major failure results.

Similarly, overload protectors are sometimes "jumped out" to stop nuisance trip outs of the system when the protector overheats. They should not be shorted out and prevented from doing their job, which is to trip out the motor at 125 percent of the nameplate current.

Improper Wiring

This generally shows up right away—but sometimes it is not until damage has occurred. Any of the following signs would lead to further checkout:

- motor drawing excessive amperage
- motor cycling on overload protection
- motor will not start, or runs below speed

ELECTRIC TESTING AS PREVENTIVE MAINTENANCE

The use of a megohm (a million ohms) meter as a testing device to evaluate the condition of the insulation on the winding of the stator in a semi-hermetic compressor is recognized as a valid predictive maintenance technique. However, there are problems. First, it is necessary to assume you are starting with a relatively dry system. If a good deal of moisture is present (as might be the case with a newly installed compressor), the reading may indicate that the motor winding insulation is in very poor condition and yet it may be perfectly satisfactory when dried out. The second problem is that the measurements being taken are only comparative. The principal value of these measurements is to offer a comparison with those taken earlier, in order to detect a deviation from the normal trend.

Megger readings should be taken the same way each time—with the same instrument, at the same time of day, by the same man, following the same procedure, and after the system has run the same length of time. An accurate record must be kept in a log.

The megohm reading of such a test will vary with temperature; therefore, the reading must be adjusted accordingly. Remember also that the winding itself is acting as a giant capacitor and it requires time to charge up. The rate of charge is also significant.

While there are many types of meggers, virtually all operate at fairly high voltage—often 500 volts or more. This is deliberate since a winding may show high resistance at low voltage, but break down at higher voltages. Generally speaking, the megger manufacturer's instructions should be followed carefully when making these tests.

WHEN THERE'S A FAILURE ...

Air conditioning stoppage, according to one of Murphy's many laws, always occurs about 5 p.m. on Friday evening or, alternately, on a holiday immediately before the day of the big meeting attended by all the "top brass."

As we have noted, the initial cause may have been electrical or mechanical—either designed in or the result of a component breakdown—or even of simple abuse of the equipment. But the net result will often be the breakup or burnout (or both) of the compressor.

Now what? The answer depends on a number of factors. If a big meeting is indeed scheduled, then budgetary considerations may well be ignored and speed may become the most important item. In such a case, a replacement compressor from a competent rebuilder or from the manufacturer is probably the quickest, most reliable solution. Together with a good bit of sweat and a lot of overtime, the crisis may be overcome without the "brass" ever becoming aware of it.

If time is less pressing and manpower availability and skill are sufficient, you may want to try rebuilding in place—*if* the compressor is an open design; *if* parts are readily available at reasonable cost; *if* you have literature that shows the manufacturer's recommendations of wear tolerances and procedures; *if* the physical location of the compressor gives enough light, reasonable protection from the elements and sufficient space in which to work; and *if* the savings to be enjoyed offset the disadvantage of having no warranty.

Considerations of experience and skill aside, it pays to understand the structure and philosophy of the industry that produces the equipment and rebuilds it.

THE MANUFACTURER

Virtually all of the leading manufacturers supply replacement compressors for current designs through their parts' deparment or through their distributors. Obviously, stocks are not always local or complete, but virtually all such compressors, whether open or semi-hermetic, are rebuilt or remanufactured. (These terms are interchangeable.) They are seldom, if ever, *new* in the strict sense of the word.

All commercial and industrial compressors—new or rebuilt—are sold with a one-year warranty.

On the whole, the manufacturers (either directly or through authorized shops) do an entirely satisfactory job of rebuilding or remanufacturing their own compressors.

THE REBUILDER

The rebuilder usually supplies a local or regional market with rebuilt (or remanufactured) compressors of all manufacturers (both current and so-called obsolete models). He normally maintains a large inventory. The competent rebuilder is likely to be a member of his national trade association, the Refrigeration Compressor Rebuilders Association (RCRA). He will usually welcome visitors. He is often also in the parts business and able to supply parts "off the shelf." He rebuilds to the particular manufacturer's specifications and offers the same one-year warranty offered by the manufacturer. His pricing is often lower than the manufacturer's because of competition.

Quality rebuilding involves stripping the compressor completely down, and immersing the body in a hot tank to rid it of all traces of carbon, acid, gum, and grease. Parts that meet the manufacturer's wear tolerances are cleaned for reuse; broken or worn parts are replaced by new. If it is a burned-out semi-hermetic compressor, the motor will be rewound to equal or exceed the original specification. The compressor will be reassembled, following the manufacturer's own procedures, and tested in accord with the manufacturer's own techniques. It is then sealed and evacuated, charged with dry nitrogen, rechecked for leaks, finish-painted, and skidded. Whether from the manufacturer or competent rebuilder, it should function "like new."

SUMMARY

Many compressor failures can be avoided by using common sense and taking care to set up a regular system of inspection, checkout, and component change. The methodical monthly checkout (including an oil analysis) will, by itself, reduce compressor failures by two-thirds.

The enemies of reciprocating compressors are moisture, liquid slugging, and electrical problems.

A compressor failure—whether the result of abuse or not—brings the system to a complete standstill. How quickly that particular system has to be back on line is usually the determining factor in how the failure is corrected. If it is a critical system, the fastest, most reliable and, therefore, least expensive solution is usually a fully warrantied rebuilt or remanufactured exchange compressor.

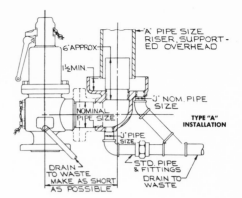

7

Operation and Maintenance of Boilers

Frederick S. Hodgdon

SCOPE OF CHAPTER

Chapter 7 covers the proper preventive maintenance of burners and boilers in order to optimize combustion efficiency and to provide proper safe operation of steam and hot water heating boilers. Specific recommendations are made for the tuning up of burners and the servicing of safety controls. Significant fuel savings can be achieved by following the recommendations in this chapter.

A boiler may be defined as a vessel in which water or any other liquid is heated and/or steam or any other vapor is generated under pressure or vacuum by the application of heat from any source such as:

1. Combustion of any fuel such as coal, oil, or gas
2. Electricity
3. Nuclear energy
4. Solar energy

Boilers may be classified as follows:

1. Process and heating boilers—low pressure
   a. For steam—operating at pressures below 15 psig
   b. For hot water—operating at pressures not to exceed 160 psig, and/or temperatures not to exceed 250°F at the boiler
2. Process and heating boilers—high pressure
   a. For steam—operating at pressures above 15 psig
   b. For hot water—operating at pressures exceeding 160 psig, and/or temperatures exceeding 250°F at the boiler

OPERATING PRESSURE

Boilers cannot be operated at their design pressure, as safety valves must be set at design pressure by codes and laws. The following suggestions are made to provide proper safety valve seating and to prevent leakage:

1. For 15 psig design steam—operate at 10 psig. Operation at 12 psig occurs in special cases.

2. For 30 psig design hot water—operate at 26 psig. This is for total of both static and velocity heads. For hot water above 30 psig, boiler design pressure should be 10 to 20 percent above the total of system static and velocity heads.

3. For high pressure—operate at least 10 to 15 percent below the design pressure. For example, operate a boiler designed for 150 psig at a pressure no higher than 135 psig.

Never start up a new boiler nor operate one with which you are unfamiliar without first checking the steam safety relief valve for:

1. Pressure setting
2. Relieving capacity

Both should match the design criteria for the boiler.

The Kunkle Valve Company, Inc., makes the following recommendations with regard to safety/relief valve installation:

1. Before installing a new valve, a pipe tap should be used to assure clean-cut and uniform threads in the vessel opening and to allow for normal hand engagement followed by a half to one turn with a wrench.

2. Avoid overtightening, as this can distort safety/relief valve seats. Remember: as the vessel and valve are heated, the heat involved will grasp the valve more firmly.

3. Avoid excessive "popping" of safety/relief valves, as even one opening can provide a means for leakage. Safety/relief valves should be operated only often enough to assure that they are in good operating order.

4. Avoid wire, cable, or chain pulls for attachment to levers that do not allow for a vertical pull. The weight of these devices should not be directed to the safety/relief valve.

5. Avoid having the operating pressure too near the safety/relief valve set pressure. A very minimum difference of five pounds or 10 percent (whichever is greater) is recommended. An even greater differential is

desirable, when possible, to assure better seat tightness and valve longevity.

6. Avoid discharge piping where its weight is carried by the safety/relief valve. Even though supported separately, changes in temperature alone can cause piping strain. It is recommended that drip pan elbows or flexible connections be used wherever possible, as shown in Figure 7-1.

The illustration below shows Kunkle discharge elbow and drip pan unit attached to a safety valve with female NPT outlet. For safety vales with flanged outlets—2" to 6"—use companion flange, short nipple and drip pan elbow, SKETCH A all same size as valve outlet 8" Elbow has integral 125# ANSI Flange.

RECOMMENDED INSTALLATION

IMPORTANT—Length of discharge piping must be kept to a minimum. For design considerations see articles, "Steam Flow Through Safety Valve Vent Pipes" by H. E. Brandmaier and M. E. Knebel (Dec. 1975) and "Analysis of Power Plant Safety and Relief Valve Vent Stacks" by G. S. Liao (Nov. 1974) available through ASME Publications.

Type "A" installation—for power boilers and unfired pressure vessel service.

Type "B" installation—for hot water boiler and generator service (160 PSIG/250° F. max.).

Type "C" installation—for low pressure steam boiler service (15 PSIG max.).

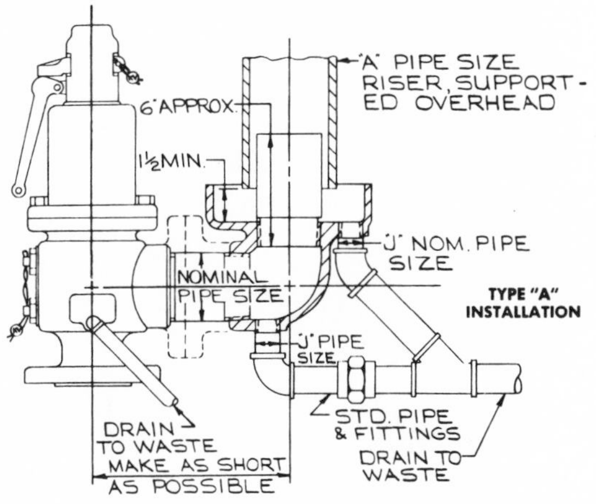

TYPE "A" INSTALLATION

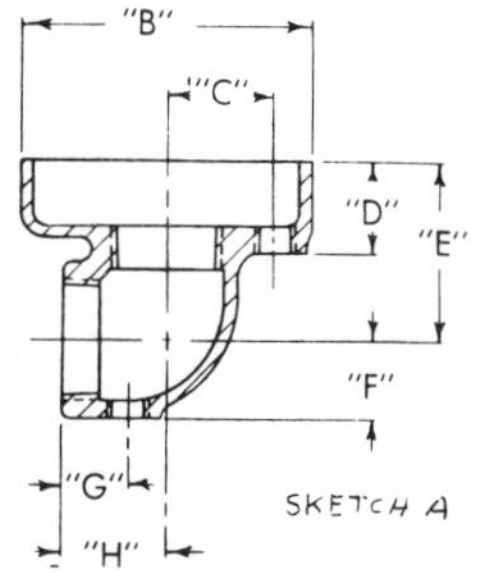

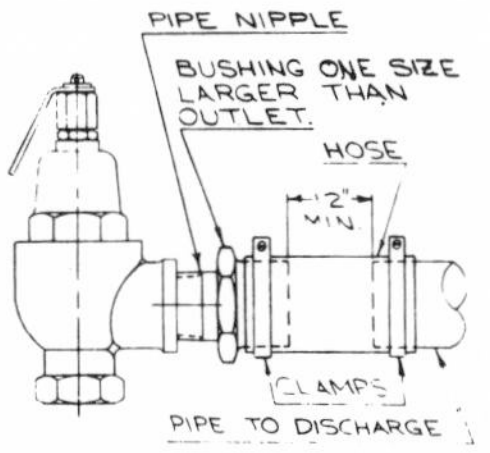

TYPE "B" INSTALLATION

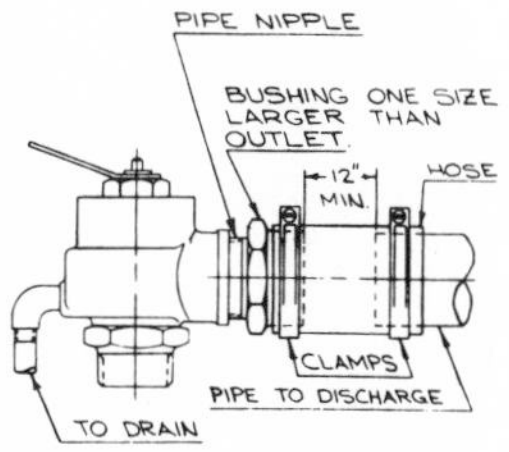

TYPE "C" INSTALLATION

DIMENSIONS

Size	A	B	C	D	E	F	G	H	J	Wt. Lbs.
2"	3	6 1/4	2 1/8	1 3/4	3 9/16	1 11/16	1 1/2	2 1/4	1/2	6
2 1/2"	3 1/2	7 3/8	2 3/4	2 3/16	4 5/16	1 15/16	1 5/8	2 11/16	3/4	10 1/2
3"	4	8	3 1/16	2 1/4	4 7/8	2 9/32	1 3/4	3 1/8	3/4	14
4"	6	9 5/8	3 3/4	2 1/2	5 3/4	2 7/8	1 3/4	3 3/4	3/4	26
6"	8	12 3/4	5 1/8	3 1/16	7 9/16	4 3/16	2 3/8	5 1/8	3/4	62
8"	10	16 1/2	6 3/8	3 3/4	9 9/16	5 3/8	7	10 3/4	1	102

Courtesy of Kunkle Valve Company, Inc.

Figure 7-1.

Drip Pan Elbows.

7. Apply a moderate amount of pipe thread compound to male threads only, leaving the first thread clean. Compound applied to female threads or used to excess can find its way into the valve, causing leakage. Flange connections should be clean and straight, with new gaskets. Draw mounting bolts down evenly.

Some safety/relief valve pointers:

1. ASME codes require that steam and air safety valves have test levers, although levers may be omitted on valves used in hazardous or toxic gas service.

2. Steam safety valves may be used for air service but *not* vice versa. Liquid valves should be used for liquids only.

3. Safety valves should be installed vertically with the drain holes open or piped to a convenient location.

4. The inlet to and the outlet from a safety valve must be at least as large as the safety valve connections.

5. Every safety/relief valve is individually tested and set by the manufacturer. Steam valves are sealed to prevent tampering. Steam and air valves may be reset, plus or minus 10 percent, with your inspector's approval. Liquid valves are adjustable plus or minus 20 percent.

STARTING A BOILER

1. Read the manufacturers' instruction manuals as many times as required to fully understand them before attempting to start up the burner and boiler.

2. Set all control switches and breakers to *off*. Check electric power supply for correct voltage and phase characteristics.

3. Check on fresh air supply to boiler room—open dampers and louvers.

4. Check on fuel supply.

5. Make sure water level shows in gauge glass. Open gauge glass valves. Use try-cocks if available to check on gauge glass water level.

6. Thoroughly ventilate boiler combustion chamber to remove any unburned fuel or gases.

7. Carefully remove and clean glass of flame scanner of combustion control unit such as a *fire-eye*.

8. Open cold water supply to boiler feed system. Open suction and discharge valves on boiler feedwater pumps. Set electrical switches on boiler feedwater system(s) to proper operating position. When feedwater pump(s) starts, check pump rotation.

9. Check pressure settings on all boiler operating and safety controls.

10. If fuel oil is used, turn on fuel oil pumps. Check fuel oil pressure. Open fuel, gas and/or oil valves.

11. Close circuit breaker or fuse disconnect switch.

12. Put all boiler emergency switches in *on* position. Reset manual reset on high pressure limit switch and low water cutoff.

13. Turn boiler control or start switch to *on/start* position. Stand clear of front of boiler, rear of boiler, and access or clean-cut doors as a precaution against explosive ignition of the fuel.

14. If a single-boiler installation, open steam stop valve. If a multiple-boiler installation, close steam stop valve. When boiler pressure is the same as the steam main pressure, slightly open boiler stop valve. If there are no unusual noises or vibrations, continue to open valve slowly until fully open.

15. Bring up pressure and temperature slowly and carefully, to observe that all controls and equipment are operating properly. Be certain that pressure operating control shuts down burner when the operating pressure control set point is reached.

   On those burners equipped with full modulation fuel controller, the operating pressure controller must return the burner to low fire before a burner shutdown. If this does not occur, check operating pressure controller and high pressure limit switch for proper settings. A high pressure limit switch setting lower than the maximum operating pressure controller setting will cause a full modulation controlled burner to shut down on high fire.

16. As soon as burner shuts off, check water column and try-cocks for water level.

17. Check safety valve for leaks. Test safety valve, using try lever.

18. Enter in log book the date and time of start-up, any irregularities observed and corrections made, time of burner shutdown after boiler reaches operating pressure, and any tests made on equipment. *Sign the log book.*

19. If any abnormal conditions occur during the burner lightoff or while the boiler is building up pressure, immediately shut down the boiler and burner. *Do not* attempt to start up again until *all* problems have been identified and corrected.

CHEMICAL CLEANING OF NEW SYSTEMS

Chemical cleaning of boilers and hot water heating systems is mandatory for all new systems to remove oils, greases, dirt, and other debris.

Steam boilers must be boiled out with an alkaline compound. This usually consists of a mixture of equal parts of:

1. Trisodium phosphate
2. Caustic soda
3. Soda ash

Use one-half pound (or 1¼ percent by weight of water in boiler) of this mixture per boiler horsepower. The following procedure is recommended:

1. Fill the boiler with water, allowing enough room for the cleaning mixture.
2. Dissolve required amount of mixture in water, one pound per gallon of water.
3. Add solution directly to boiler through top manhole or other convenient opening.
4. Close stop valve in header, remove safety valve, and do not cap or plug the opening.
5. Fire boiler without generating steam for 16 to 24 hours so that water is just below boiling point.
6. Bottom blow the boiler periodically during the cleaning.
7. Drain boiler and flush thoroughly with fresh water to remove all of the boiling-out compound. This is indicated when the rinse water is no longer pink to phenolphthalein or the pH is the same as the water used for flushing.
8. Refill the boiler to its operating level with deaerated or preheated feedwater. Add required chemical treatment, replace safety valve, open stop valve in header, and fire slowly to bring boiler gradually up to operating pressure. If boiler is connected with another boiler(s) to the same header, *do not* open stop valve until boiler is at system pressure.

CLEANING CONDENSATE RETURN SYSTEM

When a steam system is put in operation for the first time or after a long shutdown, all condensate should be run to waste for 7 to 10 days, or until it runs clear. Any oil, grease, dirt, iron corrosion products, or other debris in the condensate return system is thus prevented from returning to the boiler. Problems such as deposit corrosion of the boiler tubes, foaming, and priming will be prevented by this procedure.

During this period, the boiler will require the use of 100 percent raw water for makeup. Special attention must be given to water treatment and blowdown during this period. Water treatment should be checked every four hours or more frequently.

CLEANING HOT WATER HEATING SYSTEMS

Hot water heating systems must be thoroughly cleaned with a low-foaming, detergent-type cleaner. This will remove all oil, grease, dirt, fluxes, and other debris resulting from construction. The cleaner is normally used at a concentration of two pounds per 100 gallons of water in the system (0.25 percent by weight). The following procedure should be followed:

1. Fill system with fresh water, venting where needed.
2. Open all manual and automatic water valves to avoid bypassing any parts of the system, such as coils.
3. Dissolve cleaner in water—one pound per gallon.
4. Add cleaning solution to system by means of a bypass feeder or transfer pump.
5. Recirculate system for four to eight hours at 140°F if possible.
6. Drain and flush system thoroughly with fresh water until water drained from the system has the same pH as the fresh water.

If the system circulating pump has mechanical seals, have a spare set of seals available. Dirt dislodged during cleaning may damage the mechanical seal. Where the mechanical seals are equipped for external lubrication, the seal cavity supply water should be filtered with a 20-micron filter or a cyclone separator.

BOILER OPERATION

Modern boilers are provided with every conceivable safety device to shut down the burner automatically, should there be any malfunction of the equipment. It is the responsibility of the watch engineer or boiler room operator to check periodically on equipment operation and to test safety controls. Some of the tests should be conducted every watch. Any irregularities or repairs made during a watch should be duly noted in the Shift and Daily Log, Figure 7-2, giving exact time and complete details. Some of the items that are most critical are discussed below.

1. The water level of all boilers must be checked at the start of each watch or shift. Check gauge glasses regularly throughout the watch.
   Drain gauge glass. Water should return immediately when the drain cock is closed and the lower gauge glass valve is opened. If water returns sluggishly, blow down gauge glass until water runs clear and gauge glass fills rapidly.

BOILER SHIFT AND DAILY LOG

BOILER # _____   DATE _____

DESCRIPTION OF WORK	SHIFT #1	#2	#3
Full inspection of plant			
Read previous shift log			
Check water level			
Drain gauge glass			
Check low water controls			
Check high water controls			
Blowdown columns			
Check surface blowdown			
Bottom blowdown			
Check flame-out control			
Check boiler water treatment			
Fuel oil temperature			
Fuel oil pressure			
Atomizing air or steam pressure			
Fuel oil in tank—gallons			
Steam header pressure			
Feedwater temperature			
Feedwater pressure			
Raw water meter—gallons			
Stack flue gas temperature			
Smoke reading			
Check and recycle burner			
Check safety valves for leaks			
Burner changed or cleaned			
Oil and water strainers cleaned			
Fuel oil delivery—gallons			
Fuel oil burned—gallons			
Pounds of steam generated			
Remarks (use reverse side if needed)			

Figure 7-2.

Shift and Daily Log.

Repair any leaks around water gauge glass or fittings at once. Steam leaks can give false water level readings and can damage the fittings.

If water disappears from gauge glass, blow down gauge glass and see if water reappears. If water level does not reappear at once, *Shut down burner immediately. Do not turn on boiler feedwater. Do not open safety valve. Let boiler cool completely before adding water.* Do not operate boiler until the cause of the low water has been corrected and the malfunction of the low water cutoff has been corrected.

2. Periodically check the operation of the safety valve(s). Never attempt to repair a safety valve. Always replace it with a factory tested and sealed valve.

3. Bottom blow boiler every watch. Continuous or surface blowdowns are not a substitute for proper bottom blowdowns.

   When the bottom blowdown line is equipped with a slow opening and a lever action (quick opening) valve in series, always open the lever action first. This will prevent instantaneous release of boiler pressure into blowdown line, which could cause damage to system and injury to personnel due to rupturing of piping and fittings, especially the elbows.

4. Surging or wide fluctuations of the waterline in the gauge glass may be an indication of foaming or priming. This may be caused by too high an operating water level or too high a rate of steaming caused by a sudden overloading of the boiler. Oil, dirt, rust, suspended matter, excessive chemical treatment, or lack of adequate blowdown can also cause foaming and priming.

   This condition can cause false low water burner shutdowns and rapid loss of boiler water. A series of very heavy bottom blowdowns may correct this condition. The use of antifoams can also be helpful. If the condition persists, remove the boiler from the line, let it cool completely, and then drain and wash out the boiler thoroughly.

5. Check operation of the low water cutoffs and feedwater control. Blow down control regularly in accordance with the manufacturers' instructions to remove any sediments and dirt. Blow down until the burner is shut down as a test. This should occur preferably when the burner is on low fire.

   Periodically turn off the boiler feedwater carefully while the boiler is steaming to observe the water line and to determine at what point the low water cutoff shuts down the burner. If burner fails to shut down before water level leaves sight glass, shut down burner manually.

   If the burner cutoff level is not at or slightly above the lowest permissible waterline level, determine cause and correct. When performing this test, be certain to turn on the feedwater system immediately when the burner shuts down or when the water level falls below the lowest permissible waterline level, to prevent dry-firing the boiler.

BOILER ROOM
WEEKLY MAINTENANCE CHECKLIST

1. Visually inspect the firing rate control.
2. Pilot and main fuel valves—open limit switches, make audible and visual check, check valve position indicators, check fuel meters, and check for leaks.
3. Check and lubricate all burner and damper linkages.
4. Perform flue gas analysis. Log combustion efficiency.
5. Check flame failure control by closing manual fuel supply valves for pilot and main fuel valves—check safety shutdown timing. Record in log.
6. Check flame signal strength with appropriate meter. Record in log.
7. Check all belts for tightness.
8. Check all bearing, seals, and stuffing boxes for leaks and overheating. Lubricate if necessary.
9. Check air compressors for lube oil level.
10. Check air compressor dryers.
11. Check and run emergency generators, under load if possible.
12. Check all electrical controls and contacts.
13. Check all steam traps and regulators.
14. Pop test safety valves.
15. Record all work done in the log—review all previous data and compare.

BOILER ROOM
ANNUAL MAINTENANCE CHECKLIST

1. Open boiler—wash out water side with high pressure hose and clean fire side—inspect for any unusual deposits and/or corrosion. Check for any evidence of metal fatigue or failure.
2. Remove and inspect safety and blowdown valves.
3. Remove and clean ignition electrodes and pilot assembly.
4. Check all electric motors and circuits.
5. Examine all pumps and rebuild if necessary.
6. Remove and clean burner.
7. Clean out oil heaters.
8. Change oil, clean air filters, and intake screens on the air compressors.

9. Remove low and high water cutoffs, disassemble, and clean.

10. Examine all wiring and replace any with frayed or brittle insulation.

11. Clean burner and replace nozzle if required.

12. Check fuel oil tank(s) for water and sludge—pump out and clean if required.

13. Check fuel oil supply and return lines for sludge and wax build-up—clean with solvent if required by recirculation through lines.

14. Clean fuel oil strainers and filters.

15. Perform pilot turn down and refractory hold-in tests.

16. Check high-limit pressure safety control.

17. Check pressure operating controller.

18. Check low-draft interlock.

19. Check atomizing air/steam interlock.

20. Check high and low pressure gas interlock.

21. Check high and low pressure oil interlock.

22. Check high and low oil temperature interlock.

23. Check fuel valve interlock switch.

24. Check purge switch.

25. Check burner position interlock.

26. If a rotary cup burner, check rotary cup interlock.

27. Check low-fire start interlock.

28. Check automatic change over control where dual fuels are used.

29. Thoroughly clean all connecting piping from boiler to low and high water cutoffs.

30. Check and replace, if necessary, all defective thermometers, gauges, controls, etc.

31. Check inventory of boiler, burner, and control spare parts and replenish where required.

32. Record in the log all work repairs performed.

33. Inspect and repair all refractories and baffles in boiler.

CARE OF IDLE BOILERS

If a boiler is to be out of service for more than 30 days, it should be protected by either the dry or wet method of storage during the period of nonuse. The fireside must be thoroughly cleaned of any soot or other deposits. This is to prevent acid attack on the steel tubes should the deposits become moist. The water side should be thoroughly washed and chemically cleaned if necessary.

Dry layup method should be used if there is any danger of freeze up and/or if the boiler is to be laid up for very long periods of time.

1. Drain and clean fire and water sides thoroughly.

2. Completely dry all surfaces.

3. Place lime or other desiccants in trays inside the water side of the boiler. Seal boiler tightly to exclude all moisture and air.

4. Coat fire side of tubes with a thin film of rust-inhibiting oil.

5. Open boiler door at stack end of boiler to prevent flow of moist warm air through the fire side.

6. Close all valves and disconnect fuel lines.

7. To prevent corrosion of components in control cabinet, keep control circuit energized.

Wet layup method is to be used only if the boiler is to be out of service for six months or less.

1. Drain and clean fire and water sides thoroughly.

2. Remove safety valve and flood boiler to overflowing with the hottest water possible (use deaerated water if possible). Treat water with chemical treatment so as to provide 200 ppm of phenolphthalein alkalinity and 100 ppm of sulfite, minimum.

3. Check all boiler connections and correct any leaks.

4. Take a small sample of boiler water weekly to be certain that chemical treatment has not been dissipated.

When starting up a boiler that has been laid-up according to either of the above procedures, follow the procedure for starting a boiler.

WATER TREATMENT

Water side maintenance of boilers is of major importance in preventing scale deposits and corrosion (pitting). Scale formation will seriously interfere with heat transfer, which will increase fuel consumption and may in some cases seriously damage the rear tube sheet of fire tube boilers. Furthermore, scale formation on the water side of water tube boilers can so impair heat transfer as to cause loss of water circulation with resultant tube failures due to blistering or outright rupturing. Corrosion (pitting) can shorten tube life and require retubing. Since the number of times a firetube boiler can be retubed is limited, frequent retubing can shorten boiler life.

BOILER REPORT

LANE REFRIGERATION CO., INC. 3916 Long Beach Road, Island Park, N.Y. 11558 • (516) 431-0900

Name: SNCH
Address: OCEANSIDE

Week of: 2/3/80
Engineer: CF

Boiler or System #	BOILER #1							CONDENSATE						
Date	2/3	2/4	2/5	2/6	2/7	2/8	2/9	2/3	2/4	2/5	2/6	2/7	2/8	2/9
pH	11.2		11.4	11.0	10.4	10.8	11.2	6.5	9	7.3	8.0	8.0	8.0	7.5
Sulfite	60		55	70	20	35	55							
Organic Inhib.														
Nitrite		O												
Raw Water Hardness	6	F	7	7	7	8	7							
Raw Water Chlorides	50	F	45	52	15	30	42	0	0	0	0	0	2	0
Boiler Chlorides	8.3	L	6.4	7.5	2.1	3.8	6.0							
Cycles of Conc.	20	1	20	20	20	20	20							
Kemcolloid Dosage		N												
Kemcor Dosage	1	E	2	2	2	2	2							
Smoke	1-2		0	0	0	0	0							
CO₂	9		9	9	9	9	9							
Stack Temp.	488		470	480	490	475	480							
Feedwater Temp.	230		230	230	230	230	230							
Fuel Oil Additive														
Water Meter	9977194	10205?3	10048678	10163024	10078610	10097035	11070082							

Figure 7-3.

Chemical Analysis of a Boiler.

There are many effective water treatment programs available. It is not the purpose of this presentation to discuss the various types of chemicals and programs that can be used. It is not the effectiveness nor the need for water treatment that is in question, but rather the application and control of a water treatment program required to make it effective. See Figure 7-3.

No water treatment program for boilers can be effective without testing at least daily and without providing proper blowdown to control cycles of concentration and/or dissolved solids.

Boilers with significant amounts of raw water makeup must have chemical feed and cycles of concentration (blowdown) controlled in proportion to the raw water makeup rate. In addition, boilers that do not have deaerators in the feedwater system must have the oxygen scavenger fed in proportion to the boiler feedwater rate.

Figure 7-4 shows a typical automatic water treatment system that will proportion chemical feed and blowdown to the raw water makeup. Chemical feed to a boiler feedwater storage tank or to the storage section of a deaerator can be used only with a nonphosphate, noncaustic program. If a phosphate-caustic program is used, it must be fed either into the feedwater line to the boiler or into the boiler drum below the water level. Furthermore, oxygen scavengers such as sodium sulfite cannot be fed into the storage section of recirculating-type deaerators as the chemical will be dissipated in the mixing section.

CORROSION CONTROL

Corrosion of any hydronic system occurs because of the *corrosion triangle,* which consists of water, a metal, and oxygen. All three must be present for corrosion to occur. Eliminate any one and corrosion stops.

Since water cannot be eliminated conveniently from most boiler systems, corrosion can only be stopped by:

1. Effectively coating the metal surfaces with a microscopic protective chemical film. In a sense the metal is eliminated.

2. Eliminating oxygen from the water.

Corrosion inhibitors in recirculating water systems provide the protective film required to prevent corrosion. The type of corrosion inhibitor used depends upon the following:

1. The amount of raw makeup water used

2. Temperature of the recirculated water

3. Environmental Protection Agency (EPA) requirements

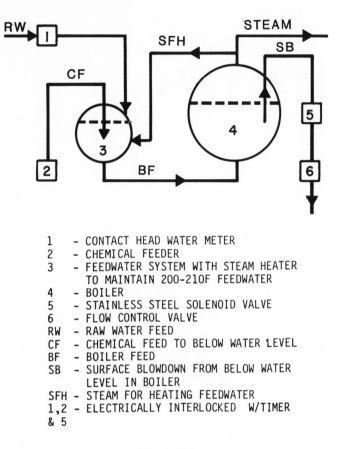

```
1   - CONTACT HEAD WATER METER
2   - CHEMICAL FEEDER
3   - FEEDWATER SYSTEM WITH STEAM HEATER
      TO MAINTAIN 200-210F FEEDWATER
4   - BOILER
5   - STAINLESS STEEL SOLENOID VALVE
6   - FLOW CONTROL VALVE
RW  - RAW WATER FEED
CF  - CHEMICAL FEED TO BELOW WATER LEVEL
BF  - BOILER FEED
SB  - SURFACE BLOWDOWN FROM BELOW WATER
      LEVEL IN BOILER
SFH - STEAM FOR HEATING FEEDWATER
1,2 - ELECTRICALLY INTERLOCKED  W/TIMER
& 5
```

Figure 7-4.

Automatic Boiler Treatment System.

Low pressure (up to 15 psig) steam heating systems and hot water heating systems (up to 250°F) can be treated with a permanent (nonoxidizable) type corrosion inhibitor that is maintained at relatively high treatment concentrations. The main advantage of such a water treatment program is that daily testing is not required and simple bypass feeders can be used. However, the system must be free from leaks and there must be no significant condensate losses. Large water losses can make the water treatment program very expensive. A water meter should be placed in the makeup line to such systems and closely monitored. When leaks occur, they must be repaired immediately.

There are many permanent-type inhibitors available, however, EPA requirements mandate the use of nonpolluting, biodegradeable, organic-type inhibitors. Chromates, on the other hand, are virtually outlawed today. A typical approved treatment would consist of benzoates, dispersants, and sequestrants to prevent scale and corrosion. Inhibitor concentration would be maintained between 2,000 and 3,000 ppm.

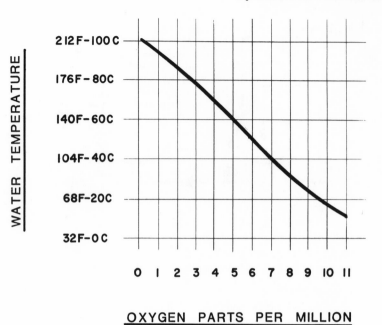

Figure 7-5.

Solubility of Oxygen in Water @ 0 psig.

Steam systems having significant amounts of makeup water because of designed water losses, such as in humidification equipment, require frequent bottom blowdown and continuous surface blowdown. The use of a permanent-type treatment becomes too costly for use in these systems.

Under such operating conditions, oxygen removal is used to prevent corrosion. Oxygen enters the boiler with the boiler feedwater. It can be seen in Figure 7-5 that the lower the water temperature, the higher the oxygen content of the water. Oxygen is removed in one of the following ways:

1. Use a feedwater heater or a deaerator.

2. Chemical treatment with an oxygen scavenger such as sodium sulfite.

3. Combined use of the above. This provides the lowest cost, and most effective method.

Chemically, about eight parts of sodium sulfite are required for every part of oxygen to be removed. Therefore, feedwater temperatures should be kept as high as possible to minimize the amount of sodium sulfite required. Feedwater pumps must be provided with sufficient suction head to prevent cavitation.

Smaller boilers should have feedwater heaters to heat the feedwater from 200 to 210°F. Larger boilers should have their feedwater provided from deaerators that have the capability to reduce the oxygen content to 0.005 ppm. Proper venting of the deaerator outside of the boiler room is necessary in order for the deaerator to operate efficiently. The most efficient deaerators are of the pressurized design, operating at 4 to 6 psig steam pressure. In addition to reducing sulfite requirements, feedwater heating reduces thermal shock to the boiler.

Dispersants are also used as part of the corrosion control program in order to keep metal surfaces clean, thus preventing deposit-type corrosion. A minimum sulfite residual of 30 ppm should be maintained in the boiler water at all times.

CONTROLLING SCALE, FOAMING, AND PRIMING

Scale is defined as the deposition of calcium, magnesium, and silica compounds on boiler tube and tube sheet surfaces. Scale must be prevented to maintain good heat transfer, thus lowering fuel costs and preventing damage to the boiler. Each 1/16 inch of scale increases fuel consumption by about 5 percent. Furthermore, scale formation on tube sheets, especially in four pass scotch marine boilers, can set up thermal stresses so great as to cause failure of the rear tube sheet.

Scale formation is encountered when the scale-forming impurities present in the raw makeup water are concentrated beyond their solubility in the water within the boiler. Scale control is accomplished by effective chemical use of sequestrants and blowdown. Regardless of the type of chemical treatment used, adequate boiler blowdown is required if scale deposition is to be prevented. The quantity of blowdown required is dependent upon the following factors:

1. Quantity of raw makeup water
2. Quality of the raw water
3. Operating pressure of boiler

Foaming is defined as the continuous formation of bubbles that have sufficiently high surface tension to remain as bubbles beyond the steam disengaging surface of the boiler water.

Priming is defined as the discharge of steam containing excessive quantities of water in suspension from the boiler, due to violent ebullition or boiling.

Foaming and priming are serious as they cause very erratic boiler operation, producing rapid loss of boiler water and an inability to maintain steam pressure. Furthermore, the water entrained in the steam can cause serious damage to steam turbines and piston steam engines. Adequate boiler water blowdown and the occasional use of antifoams are the only means by which foaming and priming can be prevented. The quantity of blowdown required is dependent upon the following factors:

1. Quantity of raw makeup water
2. Quality of raw water
3. Operating pressure of boiler
4. Design and operating characteristics of a particular boiler (boiler "personality")

The *higher* the quantity, the *poorer* the quality, and the *higher* the operating pressure, the *greater* the amount of blowdown required. When raw water is added to a boiler to replace steam and condensate losses, the impurities in the raw water are concentrated in the boiler water because they do not evaporate with the steam. There are maximum allowable concentrations of these impurities, which must not be exceeded if scaling, foaming, and priming are to be avoided.

BOILER WATER LIMITS

Boiler Pressure	Dissolved Silica	Suspended Solids (Hardness)	Total Dissolved Solids TDS
0-250 psig	125 ppm	500 ppm	3000 ppm

The above maximum allowable limits must not be exceeded. A regular blowdown schedule to control the cycles of concentration of the impurities in the boiler water must be provided. The cycles of concentration permitted for each of the above is determined by dividing the above values by the amount of each found in the raw water analysis. The total hardness in the raw water is used in the calculation related to suspended solids. Assume a makeup water has the following analysis:

Silica	10 ppm
Total hardness	100 ppm
TDS	200 ppm

Then:

$$Silica \dots\dots\dots \quad \frac{125}{10} \; = \; 12.5 \text{ permissible cycles of concentration}$$

$$**Suspended\ Solids \dots\dots \quad \frac{500}{100} \; = \; 5.0 \text{ permissible cycles of concentration}$$

$$TDS \dots\dots\dots\dots \quad \frac{3000}{200} \; = \; 15 \text{ permissible cycles of concentration}$$

**The smallest permissible cycles of concentration calculated, i.e, five cycles, must be used to establish the proper blowdown, which is calculated as follows:

$$B = \frac{M}{C}$$

where: **B** = **Blowdown in gallons**
          M = **Raw water makeup in gallons**
          C = **Maximum allowable cycles of concentration**

For example, a 100 HP boiler will evaporate about 400 gallons of water per hour at full loading. A makeup of 50 percent is 200 gallons per hour. The blowdown required is calculated as follows:

$$\text{Blowdown (B)} = \frac{M}{C} = \frac{200}{5} = \text{40 gallons per hour}$$

When either silica or suspended solids (hardness) is the factor controlling cycles of concentration, the simple chloride test is used to determine if sufficient blowdown is being provided. The chlorides in the boiler water are divided by the chlorides in the raw water to calculate the cycles of concentration being maintained.

$$\text{Cycles of Concentration} = \frac{\text{Chlorides in boiler water}}{\text{Chlorides in raw water}}$$

When total dissolved solids (TDS) are the controlling factor, a conductivity meter is used to determine if adequate blowdown is being maintained.

Blowdown can be automated as shown in Figure 7-4. The raw water meter readings can be used to determine if out of the ordinary amounts of raw water are being used. Where TDS is the controlling factor, blowdown can be automated by using a continuous sampling conductivity meter that controls a solenoid valve in the blowdown line.

CORROSION CONTROL—STEAM AND RETURN LINES

Corrosion in steam and return lines can be a major problem in steam systems. Not only is it costly to replace condensate return lines that are often buried behind walls or under concrete floors, but also the rust and other corrosion products returned from these lines and heat exchangers to the boilers by the condensate can cause deposit-type corrosion of boiler tubes. Rust also causes plugging of strainers and traps, often making a system inoperative. Furthermore, large amounts of rust suspended in the boiler water can cause serious priming of the boiler.

Steam and return line corrosion are caused by oxygen in the condensate. The corrosion is accelerated by carbon dioxide (CO_2), which makes the condensate very acidic or low in pH value. Carbon dioxide is present in the raw water and is also generated from bicarbonates and carbonates present in the raw water when it is heated in the boiler.

Large amounts of carbon dioxide can be removed from the boiler feedwater by preheating as described earlier in the removal of oxygen. Residual amounts of carbon dioxide are controlled chemically by the use of a blend of neutralizing amines. These amines are vaporized with the steam and condensed in the condensate and, in effect, neutralize the effects of the carbon dioxide by making the condensate alkaline. They also tend to film the metal surfaces, thus providing an alkaline barrier to corrosion.

Most effective results are obtained by maintaining the pH of the condensate between 7.5 and 8.5. Care should be taken not to exceed a pH of 8.5 to prevent the attack of copper and copper alloys, which may be present in the steam and condensate return system. The quantity of the neutralizing amines required is in direct proportion to makeup water quantity and the bicarbonate and carbonate content of that water.

FUEL OIL ADDITIVES

Fuel oil additives are very helpful in reducing soot and vanadium deposits on the fire side of boiler tubes. They cannot, however, be expected to eliminate soot caused by improper adjustment of oil burners. These additives contain metallo-organic compounds that are soluble in oil. Iron, manganese, and cobalt compounds lower the ignition temperature of carbon (soot) to 600°F or lower by catalytic action. Combustion catalysts in a fuel oil additive will provide significant benefits such as more complete combustion, reduced soot deposits, improved heat transfer, and a cleaner stack effluent.

Vanadium slag is a product of the fusion of various inorganic impurities in the fuel oil into a molten ash that hardens at low temperatures. Fuels high in vanadium have an ash with a very low melting point. The molten ash combines with metal tube surfaces, baffles, stack, etc., reducing heat transfer and causing serious corrosion of iron and steel surfaces. The use of magnesium and manganese compounds raises the melting point of the ash or slag to 1,400°F or higher. The resultant ash can be removed easily by soot blowing or brushing.

Fuel oil additives may also contain wetting agents, emulsifiers and organic alkaline compounds. The wetting agents and emulsifiers prevent the accumulation of moisture caused by the condensation in fuel oil storage tanks. These wetting agents will take care of up to about 2 percent of water in the oil. If large amounts of water enter a fuel tank through loose fill caps or other means, it is necessary to pump such huge amounts of water from the tank because there will not be a sufficient amount of emulsifier present to cope with the water.

Organic alkaline compounds will neutralize any acidity that may be present in water lying in the bottom of the fuel tank. Tank life is thus extended by preventing corrosion caused by acidity.

Many #4 and #6 oils have high paraffin wax concentrations. The waxes crystallize from the oil even at relatively high temperatures. Waxes have been encountered with melting points as high as 140°F. Waxes clog fuel lines, causing flame outs. Fuel oil additives with wax and sludge dispersals will prevent this from occurring.

The selection of fuel oil additives on the basis of price alone should be avoided. There are many low-priced additives that are nothing more than kerosene. In some cases such additives do not even mix with the fuel oil.

There are also some fuel oil additives, many at very high prices, that are sold on the basis of saving up to 25 percent on fuel oil usage. Nothing could be further from the truth. No fuel oil additive can substitute for proper burner adjustment. If combustion efficiency is being maintained at 78 to 80 percent there is no way that a fuel oil additive can increase combustion efficiency another 25 percent!

ENERGY CONSERVATION

It is in the national interest to conserve energy to preserve dwindling supplies of fuel. It is in the operating engineer's interest to conserve energy so as to lower operating costs. Everyone is looking for some dramatic way to reduce operating costs as fuel costs skyrocket. In most cases, however, significant savings will result only by saving a little here and a little there. The following fuel savings are possible in any typical operation:

1. Increase combustion efficiency from 75% to 80% 6.3%
2. Lower operating pressure from 125 psig to 40 psig 2.0%
3. Recover 10% more condensate 1.3%
4. Keep fire and water sides clean .... 5 %

     Total fuel savings 14.6%

Any one of the above savings is hardly a dramatic one, but, collectively, they are dramatic. Thus, if the above operation used 100,000 gallons of fuel oil per year, annual savings of 14,600 gallons of oil could be realized. At $1.75 per gallon, annual savings of $25,550 are possible!

Is the above operation a large one? Not really. It could apply to a 200 HP boiler operating at 50 percent of capacity for only 3,333 hours per year. How to obtain the above savings is discussed below.

COMBUSTION EFFICIENCY

The greatest savings in fuel can be achieved by keeping the burner properly tuned. Combustion occurs when a proper relationship exists between a fuel, oxygen (from air), and heat. Because a burner can never achieve proper combustion with the theoretical fuel/air mixture, and because heat transfer is never 100 percent efficient, it is impossible to obtain 100 percent combustion efficiency. In order to obtain combustion with little or no smoke, it is necessary to provide the combustion process with excess oxygen (air). In most cases, about 30 percent excess air is required. Combustion efficiency decreases as excess air increases. Since one of the products of combustion that can be easily measured is CO_2, the greater the amount of excess air, the lower the concentration of CO_2 will be in the stack gas. Therefore, the lower the CO_2, the lower the combustion efficiency will be.

Another indication of efficiency is the stack gas temperature. The higher the stack gas temperature, the lower the efficiency. Soot formation on the fire side acts as an insulator, preventing heat transfer from the hot gases to the boiler water. Excessive heat is lost up the stack. When using the following graphs, the stack temperature used is the *net* stack temperature. For practical purposes:

Net Stack Temperature °F = Stack Temperature °F — Boiler Room Temperature °F

Figures 7-6, 7-7, and 7-8 show the relationship between CO_2, net stack temperature, and percentage of heat lost up the stack for natural gas, #2 oil, and #6 oil respectively.

% Combustion efficiency = 100% — % Stack loss

The dotted line in Figure 7-6 shows that with 9 percent CO_2 and a net stack temperature of 400°F, the stack loss is about 17.3 percent. The combustion efficiency is therefore about 82.7 percent.

The dotted line in Figure 7-7 shows that with 9 percent CO_2 and a net stack temperature of 400°F, the stack loss is about 18.3 percent. The combustion efficiency is therefore about 81.7 percent.

The dotted line in Figure 7-8 shows that with 9 percent CO_2 and a net stack temperature of 400°F, the stack loss is about 15.2 percent. The combustion efficiency is therefore about 84.8 percent.

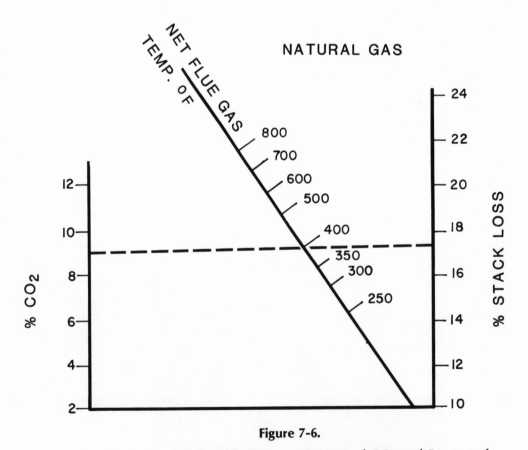

Figure 7-6.

Natural Gas—Relationship Between Percent of CO_2 and Percent of Stack Loss.

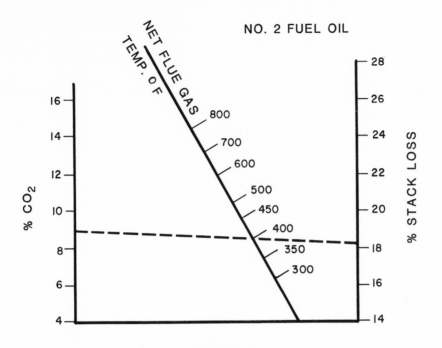

Figure 7-7.

No. 2 Fuel Oil—Relationship Between Percent of CO_2 and Percent of Stack Loss.

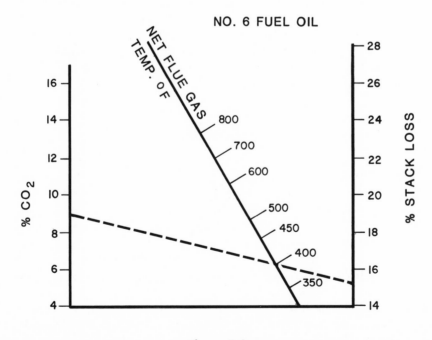

Figure 7-8.

No. 6 Fuel Oil—Relationship Between Percent of CO_2 and Percent of Stack Loss.

It can be seen from the above that with the same CO_2 and net stack temperature readings, stack losses will vary depending upon the type of fuel used. If #4 fuel oil is being used, one can determine stack losses by averaging the stack losses for #2 and #6 fuel oils. In the above examples, if #4 fuel oil were being used, the combustion efficiency would be about 83.3 percent.

Before making any burner adjustments to increase combustion efficiency, be certain to clean the water and fire sides of the boiler. When adjusting the burner, make certain that:

1. Burner is supplied with fuel at its high fire design rate when burner is operating at high fire.

2. Adjust combustion air and/or fuel rate to give maximum CO_2 with zero smoke for gas and #2 oil and a maximum of two to three smoke on heavy oils as measured by the Bacharach smoke tester.

   When adjusting a gas burner, it is absolutely essential that stack gases be tested for carbon monoxide (CO). The carbon monoxide concentration must be zero.

3. Check efficiency at low and high fire. Check through entire firing range with full modulation burners.

From time to time, various "gadgets" are promoted that are touted to increase combustion efficiency by 20 to 25 percent. These relate primarily to emulsifying water with oil or adding moisture to the combustion air. Remember that water and air add no BTUs to the combustion process. When such savings are observed, they usually result from proper adjustment of the fuel-air mixture. Remember also that when flue gas analyses show 80 percent or better combustion efficiency, little can be done to improve the operation significantly.

BOILER OPERATING PRESSURE

Boiler water temperature increases as boiler water pressure increases. At 15 psig, boiler water temperature is about 250°F; whereas at 125 psig, boiler water temperature is about 355°F. Net stack temperature at high fire will be about 150°F higher than the boiler water temperature. Figures 7-6, 7-7, and 7-8 show that efficiency decreases with increasing net stack temperature.

Fuel consumption can be decreased by lowering the boiler operating pressure. Figure 7-9 shows how fuel consumption decreases as operating pressure decreases for a boiler normally operated at 125 psig. Simply stated, less heat is lost up the stack. Many industrial boilers that are operated at 125 psig during the day, could operate at a lower pressure at night during the heating season and save significant amounts of fuel. Fuel savings as high as 4½ percent are possible during those hours when a lower operating pressure is permissible.

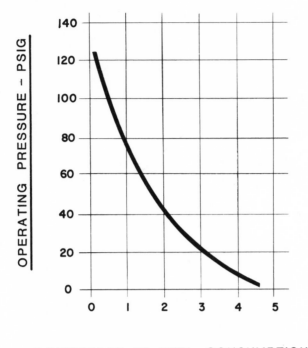

% DECREASE IN FUEL CONSUMPTION

Figure 7-9.

Decrease in Fuel Versus Operating Pressure for 125 Design Boiler.

CONDENSATE LOSSES

Another source of potentially high energy loss is condensate line leakage or other condensate losses. The loss of all returns on a 200 HP boiler operating at an average load of 50 percent means a loss of about 3,450 pounds of condensate per hour. Assuming that this condensate would have been returned at 200°F, and that the raw water used to replace it would be at 50°F, 3,450 pounds of raw water would have to be heated 150°F. The BTUs required are as follows:

$$3,450 \times 150 = 517,500 \text{ BTUs per hour}$$

Assuming the net BTUs from a gallon of fuel oil at 140,000 BTUs per gallon:

$$\frac{517,500}{140,000} = 3.7 \text{ gallons of fuel oil per hour}$$

Since at 100 HP operating load the boiler will be using about 30 gallons of oil per hour, 3.7 gallons of oil represents about 12.3 percent of the oil required for operation. Thus, every 10 percent of condensate that can be saved will save about 1.25 percent of fuel requirements. Every boiler feedwater system should be equipped with a water meter in the raw water makeup line. Condensate losses can then be measured and steps taken to minimize them. Daily water meter readings should be logged and any significant increases in raw water requirements should be investigated and corrected.

8

Electrical Systems Maintenance That Saves Money

Jack Gordon, P.E.

SCOPE OF CHAPTER

Commerce and industry depend upon reliable and adequate utility supply to function. This chapter focuses on essential elements of electric systems and building supplies, addressing and identifying those methods that optimize electric system performance, longevity, efficiency, and service reliability. These elements include the following:

1. Operating methodology and elements of good design.
2. Preventive maintenance programs to ensure reliability.
3. Breakdown response to system and subsystem failures:
   - Preparation and training assure minimum downtime and best approach to repair the failure and avoid repetition of the problem.

Electrical systems that will be examined here include:

1. Customer owned substations
2. Load centers and motor control centers
3. Motors and motor starter controllers

When utility services are well run and maintained, a reputation for performance, dependability, and organizational competence is sure to result. In addition, a measure of energy conservation will also result because clean, lubricated, and properly operating equipment will be more efficient.

General Category	Commentary	Organization
I. Maintenance facility	1. Office 2. Shops 3. Warehousing 4. Mini-computer	■ Clerk, supervisors' desks, files ■ Repair, test, overhaul ■ Inventory, shipping and receiving, equipment storage ■ Load control system, status, energy information, data-log, cost of electricity
II. Documentation	Adequate access to information via files or storage systems is mandatory.	■ Standards, codes, rules—current editions of most frequently used should be accessible. ■ Drawings, specifications and diagrams—as-is revisions to be maintained and properly filed.
III. Courses, reference material	1. Maintain a schedule of important courses given in current year. 2. Review need for technical magazines, reference books, tables.	■ All personnel should attend a minimum of one course per year to improve competency. ■ Reference library with tables of ratings, allowable levels, N.E.C. stipulations, etc.
IV. Housekeeping	Importance of shop equipment and apparatus cleanliness	■ In-house courses, lectures, films, etc., given on the value and reasons for scheduled cleaning as part of PM programs
V. Diagnostics	Requires operating know-how, proper use of diagrams, and good inventory of test devices	■ Tables of normal versus abnormal conditions ■ Symptoms of malfunctions and their interpretations in chart form are extremely useful.

General Category	Commentary	Organization
VI. Access to Data	Logical data storage and data access system is required.	■ Good filing systems require careful attention. ■ Mini-computer or data terminal with visual and hard-copy readout is best approach.
VII. Life-Safety	Safety measures take priority over all other aspects of maintainability.	■ Personnel education in safety and life-support techniques ■ Procedural techniques for on-line and off-line switching, rigidly adhered to ■ Insurances against accidental energizing, back-feeds, application of dangerous test voltages
VIII. Test devices and Apparatus	Accurate file—records inventory are to be maintained on all apparatus.	■ Instruction and parts manuals library ■ Quality assurance program ■ Training personnel in use of more sophisticated apparatus
IX. Standards	1. Installation standards 2. Materials standards 3. Forms, tags, notices standards	■ All standards should be constantly evaluated for recommendations, improvements, updates.
X. Repair and Maintenance contractors, supplies agencies, etc.	Quick access to help when needed	■ Frequently update to maintain lists of organizations and key personnel and phone numbers for emergency and urgent assistance.

Figure 8–1.

Summary of Electrical Maintenance Requirements and Practices.

247

Total electrical equipment maintenance and the factors governing each subsystem encompass a broad spectrum. This chapter selects those elements of the spectrum that are not commonly illuminated in most maintenance treatises, and deals with them in a concise manner.

GOOD MAINTENANCE PRACTICES

The maintenance or facilities supervisor must be knowledgeable and capable of planning and implementing good long-term preventive maintenance programs. His ability to convince his company that annual and long-range budgets must be available for such PM requirements is an essential prerequisite.

Some of the important requirements of a good electrical maintenance effort are listed in Figure 8-1.

An in-house skills inventory of the electrical maintenance crew should be reviewed annually to evaluate manpower capability. The facilities manager should realistically assess:

1. In-house personnel capabilities and training requirements
2. Skills and projects to be performed by outside agents and organizations

Courtesy of Westinghouse Electric Corporation.

Figure 8-2.

Typical 1500 KVA Substation (Front View).

SUBSTATIONS

Substations are systems whose purpose is to accept the utility company's primary power and transform it to a useable level. The complexity of switching, degree of redundancy, and sophistication of protective apparatus may vary over a wide range. The need for good preventive maintenance is essential if there is to be reasonable assurance of a reliable and efficient power source.

The illustrations in Figures 8-2, 8-3, and 8-4 reveal that central outdoor stations may include physical structures to support lines, switching fuses, arrestors, and grounding devices. Since the variations in subsystems are extensive, only major components are treated herein.

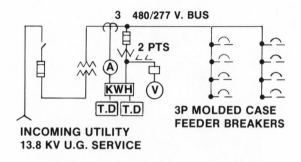

Figure 8-3.

Typical 1-Line Diagram for Above 1500 KVA Substation.

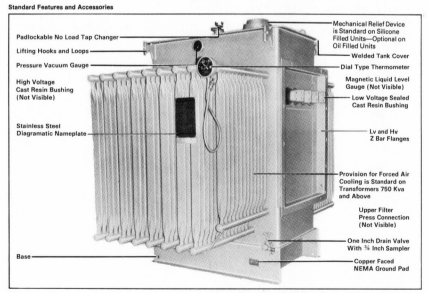

Courtesy of Westinghouse Electric Corporation.

Figure 8-4.

2000/2500 KVA OA/FA Oil-Filled Power Transformer—13.8KV/480 Volts.

Figure 8-5.
Large Outdoor 25,000 KVA Oil-Filled Transformer.

Substations can range in size from small pad mounted units of 500 to 1,500 KVA with simple primary and secondary fused disconnects; to major power transformers of 25,000 KVA and larger systems equipped with oil circuit breakers, switching structures, complex relaying, battery systems, and metal-clad secondary switchgear. Typical primary delivery voltages are 69 KV, 115 KV, and 138 KV.

A substation record and documentation library is essential and should include (but not be limited to):

1. An as-built drawing file including electrical, physical, and structural drawings

2. Specification data such as potential short circuit current values; fuse-sizing criteria and relay trip points plotted on graphic charts; backup system characteristics; breakers, and other automatic devices, trip/close timing, and clearing requirements

3. Manufacturer's manuals on the operation, maintenance, and spare parts

4. A history of prior maintenance, breakdown, overhauls, calibrations, tests, etc.

5. A spare parts list and shop capability for quick repairs when shutdown is critical

6. Addresses and phone numbers of key personnel, manufacturers, and outside technicians (A coordinated schedule of standby personnel for essential services should be readily available). See Figures 8-5, 8-6, 8-7 for typical transformer substations.

Comments on the Substation

The primary input serves to provide switching, short circuit protection, and backup to downstream protective devices. The substation structure is required to house arrestors, interlocks, p.t.'s and c.t.'s, metering relays, etc., which should be documented by a one-line diagram and component identification. The two most

Courtesy of General Electric Company

Figure 8-6.

Indoor Liquid-Filled 1000 KVA Unit Substation with Motor Control Center.

Courtesy of Westinghouse Electric Corporation.

Figure 8-7.

1500/2000 KVA Dry-Type Substation with 480 Volt Switchgear.

common voltage classifications of customer owned substations are 5 KV and 15 KV. General maintenance and repair are similar.

Transformers may vary in class, size, coolant, and configuration and yet have many similar maintenance requirements. Checking for damage, cleanliness, peeling paint, broken components, and tightness of connections may be performed informally, but a log book for recording load and voltage data, temperature levels, tap and TCUL settings, and regulator step count should be strictly maintained. A diagnostic program should be established to develop profile data against which changes in contact resistances, grounding integrity, coolant, and insulation quality can be evaluated.

Secondary switchgear will require not only testing and examination for function, but also testing for calibration and performance.

Because of complexity, the programs and recommended practices mentioned herein will be general in nature and necessarily limited to brief descriptions. More extensive descriptions of apparatus maintenance are available via the manufacturers' bulletins and publications. Complete dissertations on test methodology, test limits and durations, safe practices in handling testing equipment, and specific performance limitations, etc., can be obtained from readily available ANSI documents, test apparatus manufacturers' manuals, and other manufacturers' reference data.

A company engaged in preventive maintenance of electrical apparatus must have a maintenance staff competent in equipment maintenance shutdown, safety, and restoration procedures. Reasonable precautions, of course, must be taken to prevent accidents to personnel and equipment.

On-going training for maintenance personnel is essential to assure a competent, knowledgeable staff with good hands-on capability. Excellent programs are offered by companies such as General Electric, Multi-Amp, J. G. Biddle, and others. Test apparatus may be obtained through rental to assure that such equipment will always be operational, properly calibrated and complete with required charts, ink, etc.

Transformer Preventive Maintenance

The extent of the maintenance programs and related test and overhaul schedules will vary subject to system size, complexity, manpower availability, funding, and desired reliability level. As such, each maintenance superintendent must plan and execute his PM programs as conditions dictate.

Figure 8-8 provides a basic checklist for good transformer maintenance.

CHECKLIST FOR TRANSFORMER INSPECTION

Item	Parameters	Comment/Frequency
1. Load	1. Load Current 2. Primary, second voltage 3. KVA, KW	■ Daily nominal/maximum readings may be taken manually ■ Data acquisition/on-line transmission to central mini-computer is preferred via 15 minutes pulsed intervals
2. Temperature	1. Ambient 2. Winding/hot spot 3. Coolant	Same as item 1/comment 1
3. Gas pressure	1. Nitrogen head over liquid coolant to be maintained	■ Weekly visual inspections for manufacturer's specified minimum/maximum range; typical level from minus 2 psi to plus 8 psi
	2. Low-pressure alarm circuit	■ Nitrogen bottle pressure minimum at 2000 psi for replacement

Figure 8-8.

Checklist for Transformer Inspection.

Item	Parameters	Comment/Frequency
4. Liquid level	1. Sight glass, or bushing indicator	■ Monthly visual examination
5. Cooling equipment	1. Fans, bearings and fan motors	■ Semiannual cleaning and lube
	2. Heat exchangers (cooling fins)	■ Semiannual cleaning and check for heat emission
(Transformer out of service)	3. Dry type transformer cavity above coils	■ Annual vacuuming and check for heat flow
6. Oil leakage, rust	■ Tank, fittings, valves, gaskets, plates, etc.	■ Monthly visual inspection for evidence of leaking or rust buildup
7. Integrity of parts	1. Bushings, arrestors, gauging, instruments	■ Annual inspection for cleanliness, fractures ■ Three year hi-pot test
	2. Pressure relief, sudden pressure relays	■ Annual inspection for general condition; manufacturer's recommendation for functional test
	3. Ground connections, clean fittings, free of rust	■ Annual inspections; ducter test for continuity
	4. Controls, relays, alarms, gas analysis	■ Annual inspection; functional test; calibrations (two years) where feasible
8. Insulating liquid	1. Dielectric strength 2. Acidity level, color 3. Surface tension 4. Moisture, power factor	■ Quarterly or semiannual oil sample analysis
9. Physical inspection	1. Arrestors 2. Insulators, bushings 3. Cooling fins, heat exchanger apparatus	■ Annual inspection for cracks, chips, dirt, leaks

Figure 8-8 (*continued*).

Item	Parameters	Comment/Frequency
	4. Tap changers	▪ 24 month or 36 month operational check and calibration; ductor test for continuity of contacts
10. Electrical tests	1. Insulation resistance	▪ Annual megger testing comparative values are essential
	2. Apparatus dielectric Tests	▪ AC hi-pot testing should be performed by qualified testing laboratory
	3. Contact resistance	▪ 24 month or 36 month ductor test program

Figure 8-8 *(Continued).*

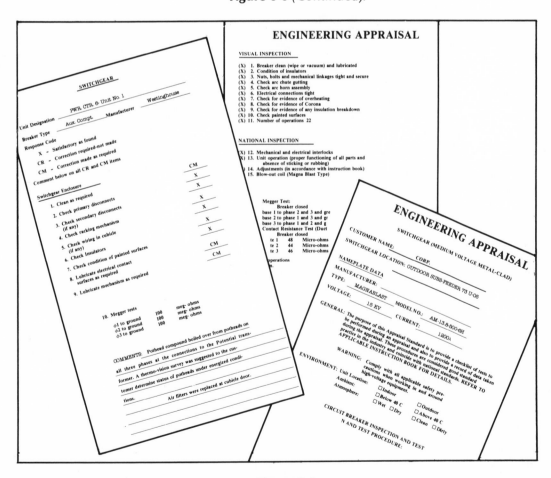

Figure 8-9.

Typical Documentation and Forms Used for Inspection Records.

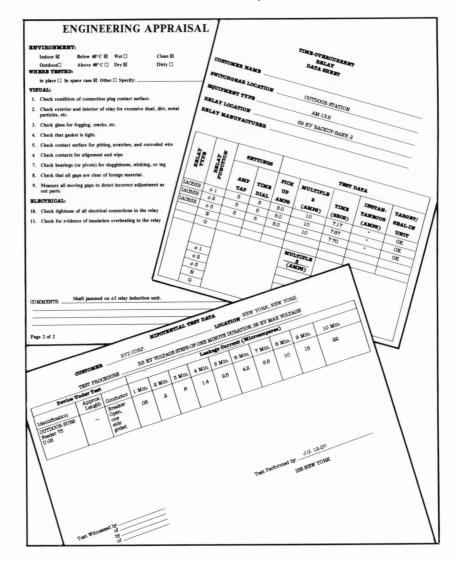

Figure 8-10.

Typical Records for Electrical Maintenance Programs.

EQUIPMENT MAINTENANCE PROGRAMS

General Introduction

An effective maintenance program is mandated by the need to keep high cost apparatus reliably functional as long as possible. Electrical equipment can remain in service 30 to 40 years when well maintained. Obsolete parts or inadequacy at meeting the requirements of system growth should be the only reasons for replacement. The value of a good program will manifest itself in numerous ways, which include longer life, dependability to operate effectively under stress and emergency conditions, and reduction in failure. Essential aids to permit a continuity of maintenance are briefly noted below in Figure 8-11.

1. Records	■ An inventory of all apparatus including catalog or order numbers, manufacturer's name, and date of manufacture. Typical Record and Inspection Forms are shown in Figure 8-9.
	■ Maintenance requirements with schedules and work to be performed
	■ History of PM and repairs
	■ A good library of drawings, parts lists, etc.
	■ Up-to-date spare parts inventory record system
2. Test	■ Reports, data sheets, and related information. Typical Maintenance Data Sheets are shown in Figure 8-10.
	■ A description of shutdown, restoration, and safety sequence practices
	■ Basic equipment such as a breaker test cabinet, grounders, closing device, couplers, relay test plugs, oil test kit, etc.
	■ Spare replacement parts inventory, as recommended by manufacturer
3. Tools and Instruments	■ Adequate full-time power supply and test benches outfitted with receptacles, ground studs, overhead hoist, factory air (clean, filtered 80-100 psi)
	■ Ammeters, voltmeters, ohmmeters, instrument transformers; megohmmeters, oil tester
	■ Filter press
	■ Compressor for blowout
	■ Hoists
	■ Vacuum pump

Figure 8-11.

Aids to Permit Continuity of Electrical Maintenance.

Checklist for Major Substation Components

Section 1.   Power Switches and Disconnects
Section 2.   Transformers
Section 3.   Regulators
Section 4.   4KV and 13KV Air Circuit Breakers
Section 5.   Equipment Rooms/Substations

Section 1. Power Switches and Disconnects

24 Month Inspection Program: System de-energized

1. Check, replace if defective, bolts, insulators, horns, shunts, linkages, etc.

2. Clean insulators, blades, connectors, contacts, hardware, etc.

3. Check alignments and stops as per manufacturer's instructions.

4. Lubricate. Degrease, prime, and paint if required.

5. Check for tightness with torque wrench, all bolts, connectors, and ground connections.

6. Megger and ducter tests. Grounding switches ducter tested to neutral with blades closed.

7. Motor operators and automatic throwover:

   a. Inspect for misalignments, proper sequencing, time delays, linkage traverse, damage, scorch marks.

   b. Test mechanical stops, limit switch travel, control devices, alarms, instrumentation, seating of blades.

   c. Check and calibrate control voltages, functions, sensing, sequencing, etc.

   d. Test auxiliary heaters, fans, interlocks, lights, receptacles, PB stations.

8. Power Fuses

   a. Clean, inspect contacts, insulators, fuse holders.

   b. Check for damage, misalignment, correct size, and pitting.

   c. Clean, tighten, and test all connectors.

Section 2. Transformers

1. Indoor dry-type units with *Nema 1* sheet metal enclosures should be inspected and cleaned annually (semiannually in a dirty environment). After disconnection and grounding, covers, and plates are removed to permit examination for dirt, deterioration, loose connections, discolorations, etc. All surfaces and compartments may be vacuumed and wiped clean with a dry cloth.

Where moisture incursion is suspected, which may occur in damp environments on unloaded transformers, units may be slowly dried by generating self-heat. This is usually accomplished by short-circuiting the secondary terminals and supplying 3ϕ low voltage power to the primary windings of such magnitude as to permit secondary flow of 50 to 75 percent of full load current. For 4.16 KV primary windings, 120/208 volts may be adequate; however, a variac should be used to control input voltage. Additionally, winding temperature must be monitored and maintained carefully at a 50 to 80°C range. The transformer may be checked for insulation resistance in 48 hours and if satisfactory, put back in service.

2. Oil or liquid filled transformers maintain their integrity by being topped with a column of dry, pressurized nitrogen in a sealed tank. The penetration of moisture and dirt due to poor seals or punctures degrades both the insulating quality and cooling ability of the liquid. Sludging is accelerated and the transformer life is greatly reduced. A major element of transformer maintenance and deterrent to failure is oil or coolant sampling and analysis for moisture and dielectric quality.

   It is essential to determine whether the coolant is mineral oil or one of the askarels (such as pyranol or interteen), which are no longer furnished due to prohibition by EPA. Test methods and results of the two liquid categories are slightly different; additionally, askarel containment to prevent leakage has become so critical under EPA law enforcement that the user is encouraged to replace such liquids with substitutes that are acceptable for indoor use or locations near buildings. (See section on insulating liquids.)

3. Transformer Diagnostic Summary

Problem	Possible Cause	Recommendations
1. Overheat	■ Coolant not functioning due to:	■ Fill as required.
	a. Low level	■ Sampling, filter press, or replace liquid.
	b. Contaminated	■ Check fan control circuit, power supply, fan motor.
	c. Fans inoperative	
	d. Fins or vents blocked	■ Clean out debris and material, check for damage.
	e. Circulating system blocked	■ Check for sludging; drain and filter liquid as per manufacturer; detank and clean as per manufacturer if above is ineffective.
	■ Overload	■ Load reduction; cooling fans or pumps switched on to manual.
	■ Overexcitation	■ Reduce primary input voltage.
	■ Winding short circuit	■ Ratio test; core loss test and inrush. If out of specification return to manufacturer.
2. Oil contaminated or unsatisfactory	■ Low dielectric strength Moisture or dirt incursion due to leaks, broken relief diaphragm, loss of pressurehead, broken gasketing, seals	■ Filter press oil. ■ Physically check and replace as indicated.
	■ Carbonization, varnish, sludge, discoloration, high acid	■ Re-refining process by transformer oil consultants. ■ Detank, wash core and coils.
3. Physical failures a. Winding open	■ Insulation or oil failure ■ Short circuit or overload	■ Manufacturer to recommend overhaul and test.

260

Problem	Possible Cause	Recommendations
b. Core failure	■ Core insulation breakdown ■ Core short-circuiting ■ Core open or loose	■ Manufacturer to recommend overhaul and test.
c. Bushing failure	■ Cracks or discoloration due to vandalism, lightning	■ Replace bushing. ■ Check arrestors and grounding, replace if faulty.
d. Leakage of oil	■ Faulty installation of bolts, valves, filler ■ Faulty gaskets ■ Cracked weldments, holes ■ Faulty auxiliary devices, press-relief diaphragm	■ Detank, remove core, drain oil, remove faulty component, replace with new item. Clean and wash down thoroughly; remove dirt, chips, etc. ■ Weld as required. ■ Replace auxiliary devices if indicated.
e. Rust	■ Weathering, scraping and abrading, rough handling	■ Derust, prime, and paint.

4. Transformers (Central Power Units)

Scheduled P.M. Programs

Monthly

1. Read and record:
   a. Ambient and liquid temperatures
   b. Winding temperature
   c. Minimum-maximum load current
   d. Minimum-maximum voltage (Primary and Secondary)
   e. Liquid level
   f. Bottle gas pressure
   g. Transformer gas pressure

2. Inspect and correct as follows:
   a. Circuits, control and fan equipment
   b. Control voltage
   c. General cleanliness
   d. Discoloration, corrosion, damage, loose connections
   e. Arrestors and insulators for dirt, cracks, chips
   f. Oil or liquid leakage
   g. Visual inspection
   h. Tank cracks or damage

Annual

1. Test oil sample; take corrective action if deficient.

2. Cooling fans, pumps and associated controls—clean, test, adjust

3. Check auxiliary devices and cabinets for heaters, lights, receptacles, operations counters, fusing, etc.

4. Inert gas pressure system: check gauging, valves, piping; nitrogen bottle replacement as per manufacturer's recommendations

24 Month

1. Incipient fault gas test

2. Clean and adjust all operating mechanisms, load contacts, alarms, controls, instrumentation; transformer de-energized.

3. Operational test for alarm, control, relay devices, and others.

4. Replace pressure bottle.

5. Clean and check for tightness all porcelain, bushings, arrestors, terminals; transformer de-energized.

6. Check, clean, paint as required fins, radiators, bleeders, breathers, etc.

36 Month (Transformer De-energized)

1. Megger and ducter tests

2. Operational test on noload tap changers, relief devices

3. Clean, degrease, remove oil stains; prime and paint as required.

4. Check and replace, if required, connections, gaskets, bushings, indicators, etc.

5. Calibrate and check all indicating devices.

6. Check and tighten tank grounding, system grounds, line and load side connectors.

7. Detank, if required, to repair internal wiring, gaskets or to check sludging.

Section 3. Regulating Equipment

Scheduled P.M. Programs

Annual

I. Step-Type Induction Regulators

1. Oil level and test for quality, check for oil leaks.

2. Calibrate and check voltage steps, bandwidth, time delay, normal voltage.

3. Check control and auxiliary voltages.

4. Check cooling fans, pumps, heaters.

5. Inspect control cabinet, dials, auxiliaries. Clean as required.

II. Tap Changing Under Load (TCUL)

1. Test control and P.T. voltages

2. Same as I.2

3. Same as I.5

24 Month

I. Step-Type Induction Regulators

1. Operational test alarms, time delay devices, control and instrumentation circuits

2. Megger and ducter tests

3. Clean, degrease, prime and paint.

4. Check and clean bushings, gaskets, indicators; replace if necessary.

5. Calibrate gages, indicators, CMV, etc.

II. TCUL Equipment

1. Drain oil, flush out tap changer; filter oil and replace.

2. Check oil level and test oil for quality; lubricate indicator gears.

3. Refer to I.2, I.4, I.5 above.

Section 4. Indoor 4 KV and 13 KV Air Circuit Breakers

Scheduled PM Programs

Annual

1. Trip/close test from control switch and all relays.

2. Check trip and close power supply; voltage/current.

3. Visually inspect for general condition.

4. Check for lights, auxiliaries, alarm, control and instrumentation functions.

5. Check wire and ground connections; clean insulators.

6. Check drawout for alignment, continuity and reconnection.

24 Months

1. All annual items

2. Megohmmeter tests

3. Continuity (low resistance) tests

4. Derust and paint as required.

5. Clean parts, cubicles, shelter, cabinets, etc.

36 Months

1. All annual items

2. Lubrication

3. Check adjustments and tolerances.

4. Check and repair blade torque, arc chutes and barriers, solenoids, etc.

5. Relay calibration program.

Switchgear—Annual PM Program

Check	For	Recommendations
1. Overall Equipment	■ General cleanliness, damage	■ Wipe and vacuum clean. Use recommended solvent and wipe dry. ■ Check for damage, overheat, corona, deterioration of insulations.
2. Moving parts	■ Wear, alignment, discoloration, rust, stiffness, loose bolts	■ Lube lift device, hinge points. ■ Align couplers, tracks, draw out. ■ Check shutters, interlocks limit travel, auxiliary relays and switches ■ Inspect tightness of bolts, connectors, wiring, auxiliary blocks.
3. Electrical components	■ Wear, damage, loose wiring of devices, discoloration	■ Inspect strip heaters ■ Inspect arc chutes; blow out dirt. ■ Main contacts ■ Control wiring, terminal blocks, fuses, breakers
4. Oil-filled breakers	■ Maintenance of oil integrity	■ Untank, check contacts for pressure, alignment, cleanliness. ■ Dielectric oil test
5. Instruments and auxiliary transformer	■ Good condition and functionability	■ Inspect all meters, lamps, switches; auxiliary transformer clean, test and calibrate
6. Test	■ Performance	■ Test trip and close; trip from relays ■ Megger test from each terminal to case or frame. Minimum of 200-1000 megohms is acceptable.

Section 5. Equipment Rooms/Substations

Semiannual

I. Battery Room

   1. Storage batteries

      a. Visually inspect for damage, spills, corrosion.

      b. Clean, tighten connections, check bus bars, ductor test.

      c. Clean batteries, racks, room; prime and paint frame if required.

      d. Check solution level and specific gravity.

      e. Check room lighting, heating, ventilation, and receptacles.

      f. Check line voltage. Spot check cell voltage.

      g. Check alarms, instrumentation, and recorders.

   2. Battery chargers

      a. Test operation, examine, and clean.

Annual

1. Batteries: load-test to verify discharge capability.

2. Battery chargers: check elements, wiring, auxiliary devices. Test for output versus load versus harmonic content.

3. Structures and shelters

TEST APPARATUS

The following is a list of test equipment that should either be owned or available through rental.

Megohmmeter

This is an instrument used to measure insulation resistance, generally utilizing a fixed potential source such as 500 VDC or 1,000 VDC to drive a test current while measuring the insulation resistance. Megohmmeters may be either battery or generator powered. The absolute reading in megohms is not as significant as the corrected trend taken over a period of time. The absolute reading is not reliable due to temperature variations and differences of dielectric absorption time for various insulations. See Figure 8-12 for typical megohmmeters. The following factors must be considered:

A. Temperature Coefficient

Insulation resistance decreases with a rise in temperature. At a reference temperature of 104°F (40°C) readings may be corrected as follows:

$$R(\text{at } 40°C) = K (R_T)$$
$$K = \text{Correction Factor}$$
$$R_T = \text{Actually Measured Value}$$

B. Voltage Rating

The megohmmeter is not a high-potential test device. Applied DC voltage should be compatible with the system under test. A reasonable guideline is offered:

	System	*Megohmmeter*
1.	208 V	250 V or 500 V
2.	480 V	500V or 1,000 V
3.	2.4 KV	2,500 V
4.	4.16 KV	5,000 V

C. Reading

The current flow (which may be in order of nano-amps) through an insulating material will build slowly to some peak level. Readings should be timed so that buildup time is consistent and resistance values can be related to each other.

D. Moisture

Moisture is a conductive contaminant that influences resistance. Resistance readings tend to be lower at higher voltages.

A 20 to 30 percent reduction in readings taken at ratios of 3:1 can signify insulation breakdown caused by contamination.

$$\% \text{ Reduction} = 100 \times \frac{\text{Rhi V.}}{\text{RLo V.}}$$

E. Low Resistance Ohmmeter

The low resistance ohmmeter, of which the Biddle "Ducter" as illustrated in Figures 8-13a and b is an excellent example, is an extremely valuable adjunct to the test arsenal. These instruments are capable of reading contact resistance down to (1) micro-ohm and up to 10 or 20 ohms. Their primary function is to read the resistance of closed breaker, switch or relay contacts.

Poor contact continuity due to pitting, filming, dirt, moisture, contaminants, misalignment, etc., becomes quickly evident. The corrected profile of resistance vs. time is far more significant than a spot absolute reading. A rising curve indicates trouble as illustrated in Figure 8-13c.

Courtesy of James G. Biddle Co.

Figure 8-12.

Typical High Quality Hand Crank Insulation Tester Capable of Measuring Megohms.

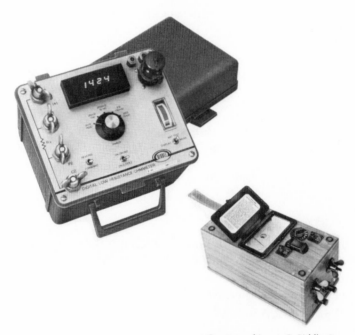

Courtesy of James G. Biddle Co.

Figure 8-13a and b.

Low Resistance Ohmmeter.

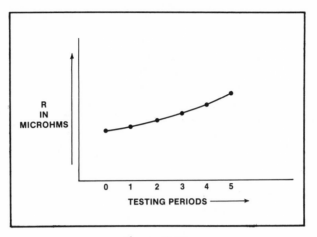

Figure 8-13c.

A typical contact profile for a 13.8 KV 1200 amp 3 pole OCB depicting a rising contact resistance.

High-Potential Testing (Hi-pot)

Hi-pot testing should only be undertaken by competent technicians who are thoroughly familiar with all aspects of nondestructive testing and the safety precautions required.

AC testing—primarily a good/bad indicator-type of test. It does not indicate the quality of the insulation under test.

DC testing—A qualitative breakdown test that measures leakage current versus time and rising potential, as per the following:

Step 1

A 10-minute test at 1/3 test potential plotting $I_{leakage}$.

A sudden increase in leakage current indicates a faulty cable; the test should be stopped immediately. Manufacturer's recommendations must be followed on safe procedure and setting of overcurrent trip to avoid cable damage. See Figure 8-14 for Typical Hi-Pot Test Characteristics.

Step 2

Increase DC potential in 10 one-minute increments to the maximum test voltage, measuring $I_{leakage}$ versus test voltage. An upward bend or knee in the plotted curve will indicate faulty insulation; the test should be stopped immediately.

Step 3

De-energize hi-pot device and apply grounding for discharge for 30-minute minimum period. After disconnection (with rubber gloves and hot stick) test again for static voltage on the apparatus or cable being tested with high-voltage detector.

269

Warning

Hi-pot testing should be undertaken only by authorized and trained electricians, following the manufacturer's recommendations. High test voltages can be lethal and/or destructive to the equipment being tested.

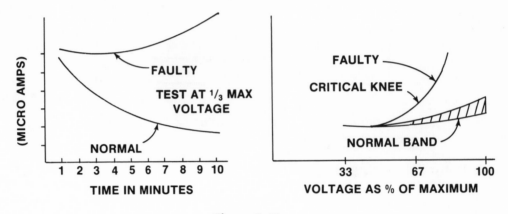

Figure 8-14.

DC Hi-Pot Test Characteristics of Insulation.

Apparatus Hi Pot

Electrical apparatus may be subjected to hi-pot testing to confirm its ability to sustain momentary overvoltages without damage or breakdown. The following treats the testing of transformers, lightning arrestors, and power capacitors. Insulators, circuit breakers, bus work, cable, and switches may be tested similarly, except that the manufacturer's recommendations for maximum applied potential during field testing should be obtained.

A. Transformers

Transformers are required to meet ANSI performance specifications for the basic impulse level (BIL). This refers to short-time high-voltage incursions such as steep-waved lightning strokes. Applied field testing is conducted at much lower voltages, approximately 1/3 of the specified BIL. The purpose of hi-pot testing of transformers is essentially to test the integrity of the insulating materials. A poorly sealed transformer will permit entry of contaminants such as moisture and oxygen, which can cause oil deterioration or sludging, and consequent reduction in insulation value. Dielectric strength of the oil can also be determined through periodic oil analysis (semiannually or quarterly). Loss of the nitrogen charge can be detected by pressure tests.

Where leakage is suspected, the transformer will require close inspection, and possible draining and disassembling. This is a detailed process with precautions taken to minimize gas bubble entrapment, and to assure proper nitrogen pressurizing. This should be performed by certified testing companies with proven ability because the oil (and transformer tank) must be desludged, filtered, and tested, to insure quality and dryness.

B. Lightning Arrestors

Generally, arrestors at small substations are of the pellet/series gap-type. Annual inspection is recommended to check for discoloration, breakage, dirt, or signs of overheating. Check connections to line and ground for condition and tightness.

At three-year intervals, a ductor test may be performed to ensure continuity to grounding bus. Such connections are typically less than 100 micro-ohms.

Work on arrestors should be preceded by total disconnection from its system and discharge of any nearby line capacitors.

C. Power Capacitors

Capacitors may be installed on feeder circuits or at substations for power factor improvement, voltage boost, surge suppression, or inductive load relief. A maintenance inspection program is advisable to check that they are functional and are not endangering the service to which they are connected.

Capacitors are sensitive to line voltage and ambient conditions, and intolerant of abuse. Most power capacitors are furnished with internal discharge resistors, but potential hazard exists even after disconnection. Capacitors should always be suspected of sustaining or holding a charge. Single-phase two-bushing capacitors may be discharged by insertion of a 10,000 ohm power resistor between terminals and clamped to ground, for not less than five minutes.

Manufacturer's instruction sheets should be consulted and followed where available.

A checklist is offered in Figure 8-15 summarizing some major capacitor failure modes and corrective recommendations. See Figure 8-16 for breakaway of typical capacitor.

INSULATION MAINTENANCE

Apparatus performance depends on the integrity of its insulation. Even with proper maintenance, insulation and insulators can deteriorate from aging and exposure to excess temperature. A record of insulation level versus time can provide a profile of the degradation process, and signal a need for overhaul or replacement.

Schedule—Minimum of two times per year

Process	Effect	Result	Program for Detection	Program for Preventive Maintenance
Excess temperature	Dehydration oxidation	Brittle; disintegrate under vibration, shock; failure	Semiannual insulation test, comparison	Thorough cleaning and lubrication

---

Note: After disconnection and discharge, monitor for voltage across terminals and from terminal to case. Proceed after assurance that no potential can be observed.

Check	For	Recommendation
1. Case	■ Physical damage ■ Overheating (discolor) ■ Leaks, punctures ■ Bulging	■ Replace; check EPA authority for disposal of PCB filled apparatus
2. Structure, site	■ Ventilation ■ Safety ■ Rust, peeling ■ Identification	■ Area kept clean ■ Isolation, fenced enclosure ■ Refurbish, paint ■ Engraved, metal nameplate(s)
3. Fuses, bleeder resistors	■ Continuity	■ Self-indicating fuses; can be checked with ohmmeter ■ Refer to manufacturer for values of resistors and schematic
4. Nominal operating conditions	■ Operating voltage, current	■ Seven day profile with RVM and RAM; should be ±10% of rated voltage and current ■ Seven day temp-profile to maintain 50-90°F if indoor
5. Insulating, bushings	■ Breakage, hairlines, cracks, dirt	■ Replace damaged bushings ■ Megohmmeter from bushing to case, 1,000 megohms minimum.

Figure 8-15.

Checklist for Capacitor Preventive Maintenance—Annual Program.

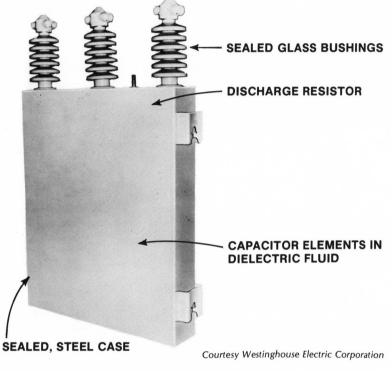

SEALED GLASS BUSHINGS

DISCHARGE RESISTOR

CAPACITOR ELEMENTS IN DIELECTRIC FLUID

SEALED, STEEL CASE

Courtesy Westinghouse Electric Corporation

Figure 8-16.

Three-Phase Capacitor Used at 240, 480, and 600 Volts.

A major objective of preventive maintenance inspection is to detect the occurrence of overtemperatures, which may reduce life expectancy and cause equipment destruction or fire.

Gauging

Temperature indicators are indispensable to effective maintenance. Oil immersed transformers generally have two indicators to measure hotspot and coil surface temperature. The dial markings provide ideal temperatures and maximum values. Pointers are magnetically resettable.

Transformer Temp/Load Record:

Four hour intervals, daily

- 8 AM to 8 PM for a two-shift plant
- Continuous for a three-shift plant

Maximum Temperature of Wire Insulations

- Based on 40°C ambient
- Life of insulation is halved for each 10°C rise of continuous operation
- Thermometer measurement

Note: See Figure 8-17 for typical limits.

Insulation Class	Maximum Temp. °C	Rise °C	Material
Class 0	90°C	50°C	Cotton, silk, paper
Class A	105°C	65°C	Same as Class 0, except impregnated, coated, or immersed
Class B	130°C	90°C	Mica, glass fiber
Class H	180°C	140°C	Silicone, mica, glass fiber with resins
Class C	220°C	180°C	Porcelain, glass, quartz
Wire Type			
RUW, T, TW	60°	—	Rubber, thermoplastic
RH, RHW, RUH, THW, THWN	75°	—	Rubber
RHH, THHN, XHHW	90°C	—	Fluorinated ethylene propylene, rubber, thermoplastic
TA, TBS, SIS, FEP		—	Crosslinked polymer, asbestos
AVA, AVL	110°C	—	Asbestos, varnished cambric

Figure 8-17.

Some Typical Insulation and Wire Temperature Limits.

Overtemperature is indicative of either overload or impairment of cooling system, which may include air/oil/forced air cooling.

Insulation Resistance

Abusive and destructive operation of electrical devices can result in insulation breakdown. The record profile of resistance readings as measured with a megohmmeter can be used to monitor this. Although minimum values of resistance to ground of [(2) + 2× line to ground voltage in KV] megohms are acceptable, most readings will be considerably higher. However, the most significant value of readings is the slope or profile of the graphic record. Readings must be temperature corrected with multipliers as noted in Figure 8-18.

Temp. during Reading		Multiplier	
°C	°F	K	
10	50	0.13	■ Reference = 40°C
20	68	0.26	■ Corrected R = R_{40}
30	86	0.50	■ Multiplier = K
40	104	1.0	■ Actual R = R
50	122	2.0	■ R_{40} = R.K.
60	140	4.0	
70	158	8.0	
80	176	16.0	
90	194	32.0	

Figure 8-18.

Insulation Resistance Multipliers.

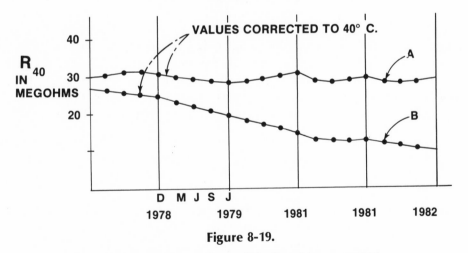

Figure 8-19.

1 HP 3ϕ 460V 60H$_z$ Motor—Stator Winding Measurement.

Curve A—Good, uniform resistance values. Slight variances may be due to temperature conversion error, moisture, or cleanliness.

Curve B—Slow degradation; the slope will probably accelerate because of insulation failure. Motor should be overhauled and checked for loading, cycling, application, etc.

Insulation Test Readings

Method	Types	D.C. Voltage Level	System Voltage
1. Megohmmeter	1. Hand-crank	1. 250 V	0-120 V
—Biddle ("megger")	2. Battery	2. 500 V	440-600 V
—Associated	3. Electronic	3. 1000 V	.6 KV-1.2 KV
—Research, Inc.		4. 2500 V	1.2K V-2.4 KV
—Hipotronics		5. 5000 V	2.4 KV-4.16 KV
—Von			
—Multi-amp			
—E.I.L., etc.			
a. Spot test	■ One-minute readings		System warm
b. Dielectric Absorption	■ Readings one minute apart for 10 minutes		" "
c. Polarization Ratio			

The dielectric absorption type of curve taken for good motor and transformer wire insulations will show a rising curve. A flat or declining curve may be symptomatic of incipient failure. Polarization indices may range from 1.0 (poor) to 2.0 (good). This is illustrated in Figure 8-19.

2. Hi-pot testing	AC	test voltage not to exceed 1.5 V_{LL}
	DC	test voltage not to exceed 1.7 V_{LL}

High potential testing is insulation testing at voltages in excess of the potential to which the system is normally subjected. Care and expertise are required to ensure safety to personnel and apparatus. System isolation and an effective "hands-off" procedure are mandated. If in doubt of such considerations, high-potential testing should be performed by a qualified outside contractor. Figure 8-20 compares the AC and DC test procedures and observed readings.

OIL

Transformers, circuit breakers, and associated devices that are filled with mineral oil require special attention to ensure that the oil, which serves to insulate, cool, and aid interruption (breakers), is functional.

Transformer coils are usually paper insulated. Immersion in the oil prevents absorption of moisture and improves insulating quality of the paper wrap. Moisture and air entrapment cause corona, oxidation, sludging, and breakdown. Circuit breaker oil interruption generates hydrogen, which must be vented.

Test	Procedure	Observation
a. AC Hi-pot	Raise potential to 1.5 × line voltage; hold for one minute and reduce to zero slowly.	Check tester for annunciation of breakdown or leakage.
b. DC Hi-pot	1. Raise potential to 0.5 × line voltage; hold for 10 minutes	1. Plot leakage current at one-minute intervals. Good insulation will show declining leakage.
	2. Increase voltage to 1.5 × line voltage in 10 additional steps at one minute; hold, per step.	2. Stop test if leakage current is rising.
		3. Record $I_{Leakage}$ versus V_{Line} on 60-second intervals. Good insulation will show uniformly rising $I_{Leakage}$.

· Note—See manufacturer's instructions on safety and operating procedures and requirements.

Figure 8-20.

Hi-Pot Test Procedure.

Dielectric Breakdown Testing

As oil ages, it deteriorates and loses its insulating quality, which in turn accelerates breakdown and even destruction of the apparatus. The dielectric test is a relatively simple procedure that can be conducted by competent on-site maintenance personnel or by testing laboratories. Oil that does not meet acceptable standards can usually be improved by circulating the oil through a filter press whose function is to remove particulate matter, sludge, moisture, air entrapment, etc.

Biddle; Associated Research, Inc.; Hipotronics, and other companies manufacture one or more testers. Excellent instruction manuals provided by these manufacturers are actually courses in oil mechanics.

Oil testing should be conducted frequently—two, three, four times a year, and always under the same conditions to ensure consistent results and produce a record profile that shows the rate of deterioration. Figure 8-21 indicates a typical apparatus for testing oil dielectric breakdown.

Courtesy of Associated Research, Inc.,
8221 N. Kimball Avenue, Skokie, IL 60076.

Figure 8-21.

Precision Insulation Oil Tester.
A Hi-Pot device is used to measure the breakdown voltage across a 0.1"
gap of oil. The sample oil is tested in the oil cup shown to the left.

Methods of Test

1. ASTM D877 standard oil test
   A measured quantity of oil sample is subjected to a calibrated voltage
   applied between two flat disc electrodes. Voltage is increased until
   breakdown potential is observed and recorded.

   a. *Test Cup*

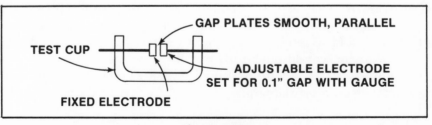

Figure 8-22.

Standard Oil Test Using Flat Gap Plates

1. Cup must be clean (as illustrated in Figure 8-22).
2. Use 111 Trichloroethane to slow down evaporation and prevent chilling.
3. Cup at room temperature.

b. *Sampling*

1. Use dry, clean quart size amber bottle with good cork.
2. Discard valve contents. Then fill bottle 1/3 full, swirl, and discard.
3. Fill bottle (almost full) and stabilize to same temperature as test cup.
4. Swirl contents and fill cup to 1/4" from top.
5. Allow oil to settle in cup for three to five minutes.

c. *Test*

1. Increase voltage at rate of 3 KV per second until current is interrupted.
2. Breakdown of 10 C mineral oil:

Application	Minimum Voltage
■ Transformer—New Oil	30 KV
■ Transformer—Used Oil	26 KV
■ OCBs—Used Oil	22 KV

3. Run five tests; average results; cycle the tests one minute on centers.

2. ASTM D1816 test
Uses spherical electrodes for more consistent, uniform testing; 0.5 KV per second rate of rise is automatic.

a. *Test Cup* (as illustrated in Figure 8-23)

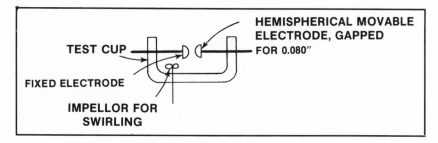

Figure 8-23.

Oil Test Using Spherical Gap Plates

b. *Test*

1. Take six tests from sample, discard first test.
2.

Application	Minimum Voltage
New Oil	32 KV
Used Oil	28 KV

3. Test is water sensitive down to 15 ppm; 40 ppm is unsafe.

Other Factors

Oxidation increases acidity and water content, which reduces heat transfer and dielectric strength. Increase in acidity accelerates deterioration, usually accompanied by darkening of oil because of absorption of contaminants.

Transformers manufactured after 1950 contain less oil per KVA of capacity and require more frequent sampling.

Oil Improvement

1. Reconditioning or mechanical filtration for removal of moisture and particulate matter.

2. Reclaiming or rerefining: heated oil (50 to 90°C) run through mechanical filter and Fuller's Earth for removal of particulates dehydration to five ppm, reduction in acidity.

3. Because of high static charge buildup during filtering, system should be extensively grounded and not removed for at least four hours after completion.

4. A summary for "Factors for Testing Oil" is provided in Figure 8-24.

ASKARELS

This is a high dielectric synthetic liquid formerly manufactured by Monsanto and used by GE as Pyranol and Westinghouse as Inerteen, among others. It was used indoors and near buildings because of its low flammability.

- No longer available because of toxic, nonbiodegradable quality.
- Very dangerous because of HCL emissions.
- Cannot be mixed with mineral oil.
- Disposal only through licensed agencies operating under EPA rules and regulations.

It is recommended that Askarel-filled transformers and apparatus be handled only by outside experts and that the earliest consideration be given to replacement of equipment or replacement of the liquid by experienced contractors in the business.

Factor	Quantify	Range
1. Acidity	Measured as Mg of KOH required to neutralize a standard sample with phenolphthaline	0.02 new oil to 1.5 bad oil
2. IFT	Interfacial tension of wire ring when pulled from oil surface	50 Dynes/cm. Excellent to 6 Dynes/cm very poor
3. Color	Ranges from new oil, very light almost colorless, to amber, and deep red to almost black	Class 1—clear white-yellow—excellent 2,3—yellow—may be marginal 4,5—orange—red—bad, to very bad 6,7—dark red to black—extremely bad
4. Water	Moisture as parts per million (ppm)	0-15 ppm—very good to good 15-30 ppm—marginal 30-40 ppm—bad
5. Sparkover	Dielectric strength in KV withstand across a measured gap	Above 30 KV—excellent 23-30 KV—acceptable Below 23 KV—not acceptable

Figure 8-24.

Factors for Testing Oil.

TRANSFORMERS

General Information

1. *Safety in Handling*

   a. Oil-filled transformers generate gases that may build up pressure internally because of inoperative pressure relief devices. Always de-energize and vent before opening or detanking. Pressure gauge should read 0 psig.

   b. Tap changers are always presumed to be no load devices. Do not operate when energized.

   c. Tank ground is permanent. Check that ground connection is solid and properly made before attempting any maintenance or examination.

   d. Transformer is normally sealed and airtight to avoid entry of moisture. A leak test may be performed by introducing five psig of dry N_2 and checking for 24 hours (corrections for ambient temperature may be required).

2. *Overheat*

   a. Transformer overheat seriously reduces life and degrades such items as oil and insulation. A 10°C rise above maximum recommended temperature for extended periods (eight hours) is said to reduce life by 50 percent.

   Adhere to the manufacturer's operating specifications and take frequent temperature and pressure readings. Immediate load reduction is advised if temperature is approaching maximum, or pressure is erratic.

   A transformer rated for 55°C rise over 30°C ambient can sustain a maximum gauge operating temperature of 85°C (185°F). Internal hot-spot temperature may rise an additional 15°C.

   b. Fans are frequently affixed to the transformer case or cooling fins, serving to boost capacity in the order of 25 percent by increasing the heat exchange from cooling fins to environment. Fan control is usually operable as manual (continuous) or automatic (temperature controlled). It is essential to ensure that the fans are functional and with unimpeded air flow.

   The manufacturer has exercised great care in fan design to ensure proper cfm of air flow and mounting to minimize vibration and resonance. Therefore, the user should not modify or alter the forced cooling system unless approved by the manufacturer.

3. *Reseal after Inspection*

   a. Transformers should be inspected internally and resealed as quickly as possible, to prevent excess moisture incursion. Oil may be dropped to core level (but not below) for physical and leak examination. If a trouble light is required, equipment should be explosion-proof and noncontaminating.

After reseal and refilling of oil, pressure with five psig of dry, Type I nitrogen and megger test the windings (1000 V test set) with oil at 68°F:

Winding Class	*Megger Range*	*Type*
5 KV	80-150	Megohms
25 KV	400-1000	Megohms
69 KV	1500-∞	Megohms

b. Low resistance measurement to case indicates a grounded coil that must be corrected before re-energizing. Frequently draining, filter pressing, and refilling of oil will clean the ground. Occasionally, high-potential testing the coil at 50 percent of its nominal rating will burn off a wire strand, causing a ground.

   A persistent, nonresponsive ground in the transformer core will require disconnection and examination by a qualified transformer repair shop.

c. Sealing technique requires the use of and following of the manufacturer's maintenance manual carefully. Gas oil, conservators, automatic gas regulator, and sealed tank with bleeder all require different handling methods.

d. Moisture extraction from oil may be assisted by high vacuum (25 mm mercury) applied to normally filled tank. A two to six hour draw will suck water from core and coil, until a measured evacuation of 1 to 1.5 ounces of water/hour is reached, indicating reasonable dryness.

SECONDARY PROTECTIVE DEVICES

It is presumed herein that the array of protective devices, including circuit breakers, associated relaying and fusing, have been properly selected, configured, and installed. Adequate, scheduled preventive maintenance must be performed, so as to offer reasonable assurance that a breaker (or fuse) will sectionalize a faulted circuit as intended.

Fuses

1. Check for correct fuse, fuse-holder type, and size, and replace if required. Improper selection of fusing may result in both loss of selective clearing *and* undersizing in ability to contain the available fault magnitude, which is extremely dangerous.

2. Fuse clips must hold a cartridge fuse firmly and tightly. Loose fuse clips can cause high-contact resistance, excess heat and subsequently frequent fuse blowing. Such clips must be replaced promptly. Frequently the fuse will be discolored at point(s) of loose contact with the fuse holder.

3. Fuses should be located so as not to be subjected to excess ambient temperature (40°C) or vibration.

4. Nonindicating cartridge fuses may be tested by checking voltage or fuse continuity.

   a. This method is positive and safe for fuse continuity. The circuit is de-energized and the suspect fuse is removed with a fuse puller and simply checked with an ohmmeter for open or short.

   b. Fused circuits may be checked by measuring voltage on the fuse(s) loadside. Zero voltage indicates one or more fuses have blown. Occasionally, a sneak circuit or path (especially in the case of three wire and three-phase systems) will give an indication, in error, of potential. This can be avoided by disconnecting all loadside feeders and taps.

   c. See Figure 8-25 for a typical test circuit.

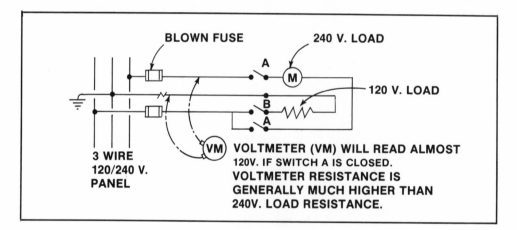

Figure 8-25.

Checking Voltage on Loadside of Suspected Blown Fuse.

Circuit Breakers

The majority of 120, 208, 240, 277, and 480-volt circuit breakers in service are molded case-type and range in size, ampacity, trip characteristics, etc., from small lighting panel use to major load use, handling several thousands of amps of sustained loading. These breakers are contained in assembled cases and are not accessible for internal examination or repair. They are frequently installed in load centers as in Figure 8-26.

Courtesy of General Electric Company

Figure 8-26.

480 Volt Switchgear.

Molded case breakers are the watchguards of industry and should not be taken for granted. Their trip and switch mechanisms are precision devices carefully calibrated to perform tripping functions within defined bands of tolerance. Additionally, the breakers may have been furnished with any of numerous modifications, such as ground-fault trip, motorized reclosing, under-voltage trip, etc. Prior to undertaking any form of PM, it is essential to review and fully understand the following:

1. Manufacturer's Technical Bulletins
2. Purchase orders describing how the breakers were furnished and their technical description
3. Time-current curves illustrating the trip characteristics

Performance of PM programs should be undertaken with competent personnel and an adequately furnished shop, which should include the normal array of tools, instruments, meters, powered benches, plus the following:

1. *Ducter* for low resistance measurement

2. *High-potential testing* to permit testing at 50 to 60 percent of specified withstand voltage

3. *Short circuit load tester* with timer, such as Multi-Amp Test Apparatus

4. *Shop tester* for the breaker to be tested (Each manufacturer has such testers available to permit bench or shop mounting for testing of the breaker.)

Key electric maintenance personnel should be enrolled annually in training centers such as those conducted by General Electric, J. Biddle, Multi-Amp, and others, where experts provide intensive shop and technical training programs. The benefits outweigh the costs of these programs. Maintenance personnel must be able to perform hands-on checkup sequences with confidence. Companies lacking the resources to justify such an investment are best advised to employ an outside agency to perform the inspections and maintenance programs outlined herein.

Recommended Test and Inspection Program

I. Breakers tripped open under short circuit action:

Such breakers that may have cleared faults in excess of 50 percent of their capability should be removed, replaced, and thoroughly bench tested and certified before returning to inventory. Breakers that may have cleared in excess of 80 percent of capability should be returned to the manufacturer for examination and recommendation to either dispose or recondition. Cost is a major factor because breakers may range in price from under $50 to over $5,000 for very large frame units.

II. Annual Inspection

a. Physically examine for damage, discoloration, distortions of adjacent bus work, loose fittings, and conductor connections. Remove and discard breaker where damage is evident.

b. Check for compartment cleanliness and reasonable ambient temperatures (20 to 30°C is ideal; up to 40°C is acceptable).

c. Drop load and operate switch on-off for several cycles checking loadside voltage. Check that operating handle has a firm action and shape. This can usually be verified after operating several breakers of the same frame and type.

III. 24-Month Inspection

Remove breaker and install on shop tester.

a. *Trip Test*
Connect short-circuit test console as per instruction and test tripping time at 150, 300, 500, and if possible, 1,000 percent of breaker rating. Test each point twice, allowing at least 10 minutes for thermal cooldown

between tests. Trip points should fall within the published minimum-maximum trip time curves.

b. *Continuity (low Resistance)*
Closed contact resistance check for good continuity. One to 10 microohms across the input/output terminals is reasonable; however, equally important is the long term profile to maintain stability. Degradation of the contact surfaces due to arcing, pitting, or overheating will cause a gradually increasing resistance. A substitute test consists of running a calibrated current (DC) through the breaker and measuring voltage across the input/output terminals with a micro-voltmeter.

c. *Insulation Resistance*
Use of a 1,000-volt megohmmeter will serve adequately to test 600-volt class of circuit breakers. See sections on *Insulation Maintenance* and *Megohmmeter* for further discussion on method and acceptable values.

Overcurrent Relays

System protection may be provided through the use of shunt trip breakers, which receive their intelligence to trip from nondirectional overcurrent relays. These, in turn, are sensing line current from the secondary output of current transformers. These relays (typically Westinghouse "CO" and General Electric type "IAC" devices) utilize the induction disc principle, whereby stator magnets with current coils induce eddycurrents into an aluminum disc. This causes the disc to turn at a speed logarithmically proportionate to the stator current magnitude. The disc, spring restrained, can be made to turn at a calculated threshold level, and close a contact at a precise time-current relationship. Thus, a breaker can be controlled to trip according to the level of short circuit, with a precision greater than that of molded case breakers furnished with thermal trip devices.

Overcurrent and other protection relays may be tested and calibrated at intervals ranging from 24 to 36 months. Additionally, their associated breakers may be trip-tested by closing the relay contacts, assuring that the relay trip circuitry is functional.

The sophistication and expertise required for comprehensive PM and calibration programs of induction-disc relays is far beyond the province of this text. Testing and engineering consultants are suggested to assist companies in preparing short circuit analyses from which relay systems may be calibrated and tested effectively.

Maintenance personnel should restrict their efforts to annual examinations for general cleanliness, wiring and connections, removal of iron filings that may accumulate around the relay magnets (after removal of the relay from its case), and checking that tap and dial settings agree with the specified recommendations.

AC MOTOR MAINTENANCE

Motors are the primary muscle on which a plant operates. These workhorses are abused through long, harsh periods of rough service with a minimum maintenance. The special classes, types, and sizes of motors are extensive; hence, only the fundamental elements of motor maintenance will be discussed. Motor controllers require the same care as load centers; see the section on motor control centers discussed elsewhere in this text.

Motor maintenance is difficult to execute because of lack of available shutdown time and because each motor type and operating environment may require widely varying inspection and maintenance procedures.

Discussion will be limited to preventive maintenance and exclude breakdown or emergency repair. It is assumed that the motor has been properly selected and installed.

I.   Major Problem Areas

A. *Dust and Dirt*

Motors with ventilating slots allow a free flow of air essential to cooling. Dust and dirt will accumulate on wire, slots, bearings, slip rings, and commutators. Generally, dirt can be blown out by applying a clean dry source of 25 to 40 psi factory air.

		Environment		
	Treatment	Clean	Moderate	Dirty
1.	Blow dirt, wipe clean	12 months	6 months	monthly
2.	Disassemble and clean	24 months	12 months	6 months
3.	Oil and grease fittings and gaps in place. Check dust seals and gaskets.	6 months	3 months	monthly

B. *Moisture*

Moisture inside the motor will cause insulation failure, corrosion, and bearing failure. Close inspection and insulation resistance testing will generally expose insipient failure caused by moisture. Drip-proof and enclosed motors frequently build up internal moisture from condensation if the motor is shut down or cycled and allowed to drop below the dew point temperature. Periodic megohmmeter test comparisons are of greater value than the actual resistance reading. (See *Insulation Resistance Testing*.)

Where permanent damage has not occurred, insulation resistance can usually be restored by application of heat, within the motor's limitations.

1. Heat Application Methods:

   a. Disassemble motor, clean with motor solvent, and place in 80°C oven for 24 hours, allowing temperature to rise slowly.

   OR

   b. Force 40°C dry air through motor and into windings. Test every eight hours.

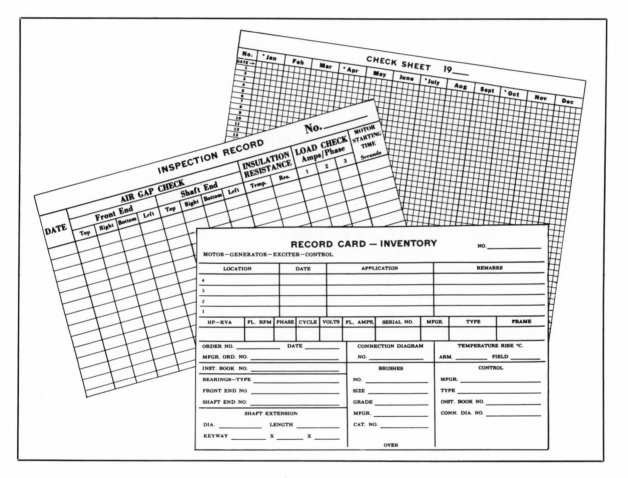

Figure 8-27.

Typical Record Cards for Motor Data and Profiles

The record card provides the historical data and evidence of scheduled maintenance and testing. An important aspect of such data is the profile of resistance and load values which may be analyzed as to sudden change and potential trouble.

OR:

   c. Apply 60 Hz. variable source low voltage (order of 3 to 10 percent of rated voltage) in the motor shop. Low current must be monitored to ensure locked rotor condition and light heating. Test every eight hours.

  2. Treatment

		Environment		
		Dry	Moderate	Wet
a.	General inspection and insulation Resistance Test.	Annual	6 months	3 months
b.	Lubricate, oil, check all plugs and fittings. Tighten.	See manufacturer's Recommendation		
		Annual	6 months	3 months
c.	Disassemble for close examination and test. Clean; bake if required.	Following poor resistance reading		

C. *Lubrication*

Motor lubrication requirements will range from oiled and greased to sealed bearings. Manufacturer's recommendations of grade, quantity, and schedule for lubrications should be followed rigorously because overlubrication can be as damaging as insufficient oil or grease. A poor quality lubricant may be high in acid content or graphite and may cause severe damage.

  1. *Oil*

Over-oiling or oiling while the motor is running may spill stray oil onto insulation, slip rings or commutators, brushes, etc.

Carbonizing or breakdown of such contamination is highly destructive and should be removed quickly.

Oil Application: a. Shut down motor.
                           b. Empty reservoir and refill.
                           c. Replace caps or covers.

  2. *Grease*

Same caution as for oil.

Application: a. Shut down motor.
                           b. Remove covers and drain plugs.
                           c. Remove as much old grease as possible.
                           d. Add new grease at low pressure until new grease emerges at drain plug.
                           e. Run motor for 10 minutes, shut down; wipe excess grease and replace covers and drain plugs.

3. *Sealed Bearings*

Replace only when indicated by overtemperature, end play, or shimmy. See manufacturer's instructions.

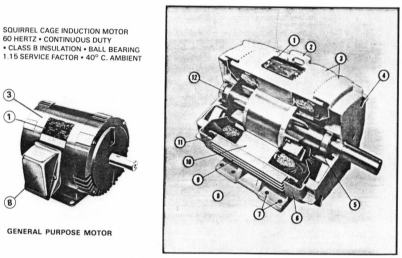

SQUIRREL CAGE INDUCTION MOTOR
60 HERTZ • CONTINUOUS DUTY
• CLASS B INSULATION • BALL BEARING
1.15 SERVICE FACTOR • 40° C. AMBIENT

GENERAL PURPOSE MOTOR

Courtesy of Gould/Electric Motor Division

Figure 8-28:

General-Purpose Induction Motor Illustration and Cross-Section.

1. Aluminum Nameplate—provides motor rating for easy reference.
2. Integral Lifting Rings.
3. Aluminum Frame and Endshields for heat transfer.
4. Grease Plugs—accessible for lubrication.
5. Endshields for maximum bearing life.
6. High quality finish.
7. Twin Bolt Holes for each foot.
8. Conduit Box—for wire connections.
   Entrance from top, bottom or either side.
9. Ribbed Construction—heat dissipation.
10. Annealed Laminations—reduce core losses.
11. Insulation System—high-temp. short-time thermal endurance; increases life.
12. Ball bearings of vacuum-degassed steel.

II.   General Motor Maintenance

Schedules for PM of motors will depend on environment, access, size, use, redundancy, application, history of problems, and, of course, whether motor is essential to plant or life safety. Economic considerations will dictate the necessity of spare motors for some applications.

An inventory record and history file of all motors is essential. A tickler system should be developed to schedule each motor for inspection and/or test. General inspection should be undertaken at least annually, or as often as monthly, where large motors operate continually in dirty, wet environments.

Figure 8-27 offers some typical inspection and checking system record forms. Other forms are available from motor manufacturers.

Photographs, drawings and cutaways, such as shown in Figure 8-28, are valuable documents and should be obtained for typical motors utilized in the plant.

A summary for a typical motor preventive maintenance program is offered in Figure 8-29.

	Environment		
	Clean	Moderate	Dirty
1. *General Inspection* 　a. Clean motor if required. 　b. Blow out dirt, dust 　c. Lubricate as per manu- 　　　facturer 　d. Check ambient 　e. Loose bolts and fittings	6-12 months	3-6 months	1-3 months
2. *Test* 　a. Insulation R 　b. Low resistance 　c. Voltage, current at 　　　zero and normal load 　d. Bearing temperature 　　　or case temperature	12-18 months	8-12 months	6-8 months
3. *Bench Inspection* 　a. Disassemble; check for 　　　moisture, bad bearings 　　　internal dirt, condi- 　　　tion of varnish 　b. Continuity test for 　　　rotor and stator 　　　resistances, grounds	3-5 years	2-3 years or as may be required Same as 3a	1-2 years

Figure 8-29.

Motor Preventive Maintenance Program.

		Environment	
	Clean	Moderate	Dirty
c. Check seals, gaskets, rings, commutators wire for cracking, splitting, color		Same as 3a	
d. Reassemble; dynamic check for speed, vibration, balance, noise, harmonics	As required	As required	As required
4. *Motors with Brushes, Slip Rings or Commutators*			
a. Check for wear, discolorations, scratches, dirt, binding, clearances	6-12 months	3-6 months	1-3 months
b. Check spring tension, brush surface, brush movement	Same	Same	Same
c. Check commutators for good color, wear, high mica, brush continuity	Same	Same	Same
d. Check for excessive sparking	Same	Same	Same

Figure 8-29 *(continued)*

III. Insulation Resistance

Acceptable resistance at 40°C for motors is derived from:

Megohms = one + general voltage in KV

A comparison of resistance vs. winding temperatures for several families of voltage levels is shown in Figure 8-30.

Large motors may be checked for dryness and cleanliness by DC insulation resistance testing close to the motor rated voltage, by obtaining its Polarization Index:

$$\text{Polarization Index} = \frac{10 \text{ Minutes test R value}}{1 \text{ Minute test R value}}$$

The apparent R of good insulation continues to rise with time, hence a high index is indicative of clean, dry insulation.

Insulation	Index	
1. Class A	1.5-3.0 1.1-1.5	Acceptable Very poor to minimum
2. Class B	2.0-3.0 1.1-2.0	Acceptable Very poor to minimum

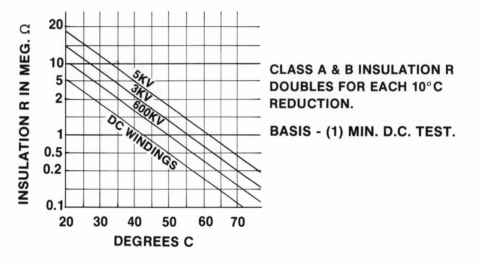

CLASS A & B INSULATION R DOUBLES FOR EACH 10°C REDUCTION.

BASIS - (1) MIN. D.C. TEST.

Figure 8-30.

Minimum Resistance vs. Winding Temperature.

IV. Motor Operating Temperature

Overtemperature causes insulation degradation. A rule of thumb advises that continuous operation for each 10°C excess, reduces insulation life by half. Hot spot temperature measurement (if embedded detector is available) is extremely valuable, because it furnishes data consistent with the class of insulation. A compromise would permit thermometer measurement of the hottest accessible part. End bearings temperature may also be measured by thermocouples or thermistors embedded in the stator slots.

Figure 8-31 provides an approximate reference for allowable temperature rise. Two assumptions are made:

1. Ambient reference is 40°C.

2. A significant difference exists between measurable values and allowable "hottest spot temperature."

Insulation	Type	Temp: Hottest Allowable	Maximum Measurement by—	
			Thermometer	Probe, TC, Resistor
Class A	Impregnated cotton, silk, paper	105°C	90°C	100°C
Class B	Mica, fiberglass, asbestos	130°C	110°C	120°C
Class H	Mica, fiberglass with special silicone binders	180°C	150°C	160°C

Temperature is based on 40°C ambient plus rise.

Figure 8-31.

Insulation Allowable Temperature Values.

V. In-Service Inspection

Valuable information is obtainable from inspections performed while a motor is operating. Schedules for such inspections depend on circumstances. Large motors operating in attended equipment or boiler rooms may be checked weekly. Inaccessible motors should be checked quarterly or at least semiannually.

Checklist

1. *Sound*

   Listen for unusual whine, rumble, vibration, or other signs of pending malfunction. An experienced maintenance technician becomes familiar with the steady, smooth sound of each motor. Loose bolts, obstructing material, dirt, worn bearings, and such will cause distinct changes in normal sound characteristics.

2. *Speed Check*

   Portable tachometers or strobetachs permit easy, accurate speed readings. Historical data will indicate if the motor is slowing down sufficiently to warrant a shutdown for causes and correction.

3. *Voltage and Current*

   A profile of voltage and line current will provide a basis to evaluate motor stability and operation. Motors operate most efficiently at optimum load and rated voltage.

4. *Summary*

A general t st summary and malfunction evidence is provided in the table of Figure 8-32.

General Test Functions		
Test for	Method	Results
1. Stator Windings short	Remove rotor use growler	Loud, growling noise; over-shorted coil
2. Stator Windings open, or ground	Ohmmeter, test continuity	Infinity = open zero to ground = ground
3. Bearing failure misalignment, loose mounting bolts	Ultra-sound detector, vibration tester	Noise must be diagnosed and compared to normal operations.
4. Insulation resistance	Megohmmeter; measure from line to case	Minimum acceptable value at one megohm plus KV rating; profile is essential.
5. Continuity	Low resistance ohmmeter	Stator windings should be within ±10 percent brush to rings or commutator minus ten 100 megohms.
6. Temperature	Embedded Sensor. Thermocouple, Thermometer, Varistor	Rise determined from 40°C ambient reference; see table under "Motor Overheat."

Figure 8-32.

Motor Tests and Malfunction Parameters.

Maintenance card form (top of page):

EQUIPMENT AND NUMBER: BUHLER MACHINE GEAR NO MOTOR.
LOCATION: 3RD FLOOR - LONG GOODS, WORK CENTER - 311
MAINTENANCE CARD - FORM #1

Column headers: GREASE FITTINGS, OIL FITTINGS, MAIN GEAR CASE, AUX GEAR CASE, MAIN MOTOR, AUX MOTOR, OPEN GEARS, CHAINS, OIL RESERVOIRS, WAYS, CRANK CASE, SLIDES, AIR LINE, HYDR SYSTEM, AUTO GREASE LUBER, AUTO OIL LUBER, CAMS

Entries: GREASE FITTINGS 3; MAIN MOTOR 26; AUX MOTOR 10; OIL RESERVOIRS 2
TC 123; TC 123 TC 123; TM 120

GEAR BOXES TO BE DRAINED, FLUSHED AND FILLED WITH TECH - LUBE BY DEC. 1979
SAMPLE SENT TO LABORATORY ON NOV. 10 1979

TECH·LUBE CORP.
127 CABOT STREET WEST BABYLON, N.Y. 11704
(516) 752 - 9000 OR (212) 527 - 6000

Lubrication Procedure – A Good Investment

by Robert Mayer

SCOPE OF CHAPTER

This chapter is designed as an aid to individuals who are responsible for preventive maintenance, including lubrication of all types of industrial equipment, heating, and ventilation found in industrial plants, and commercial and residential buildings. The necessary information is provided to set up a successful lubrication program as part of plant maintenance in a modern facility where the maintenance manager does not have a large engineering staff to develop lubrication requirements.

BENEFITS OF PROPER LUBRICATION

The goal of any planned maintenance program is to reduce costs. A proper lubrication program will increase the life of equipment, reduce failure and repair costs, minimize downtime by reducing the unplanned repairs of major equipment, and help to lower energy costs. Proper lubrication depends upon proper planning. Planning must take into account the characteristics of the equipment to be maintained. Accessibility and location of equipment, resources available in terms of manpower and their skills, and the probable time interval between failures must all be considered in order to develop an effective plan. Preventive maintenance has traditionally included the elements of inspection, lubrication, adjustment, and cleaning at regular intervals. Obviously, to plan such a program effectively, you must have on hand accurate and reliable data. It is not enough to tell a mechanic to *inspect* a certain piece of equipment. The instruction must advise him as to whether the inspection should be performed under running conditions or when the equipment is down. The instruction also should be specific in terms of whether the inspection should be visual only, or should include instrumentation such as, vibration testing, or the use of a listening device as aids in evaluating the condition of possible wear, misalignment, or lack of lubrication.

When designing the lubrication program, a decision must be made as to whether the inspection program will be limited to inspection only, or if the technician will also perform lubrication and adjustments at the time of inspection.

Inspection and adjustment should be an integral part of the lubrication program where possible. This means that maintenance instructions must tell the person who will perform the maintenance work to inspect at certain intervals when lubrication may not be normally required. Manufacturers' guidelines for inspection and lubrication frequencies and the actual needs of each individual device should be integrated in accordance with recommendations set forth in this chapter.

CLASSIFICATION OF LUBRICATION REQUIREMENTS

Most reference works classify lubrication by types of bearing surfaces such as high and low friction, high rotating speed and low rotating speed, roller, ball, and journal or anti-friction bearings, and drive mechanisms. Lubricants are classified by characteristics such as viscosity, or type, such as waxes, greases, petroleum, synthetic, or solid lubricants. While such information is absolutely essential for proper integration of good lubrication with machine design, it is of less importance to the practicing maintenance engineer who is establishing a lubrication program for his plant. To classify all the operating equipment in the plant by rotation speed, and/or lubricant base (petroleum or synthetic), would not be very beneficial in setting up a program—unless lubrication frequency and effective routes for the maintenance mechanic to follow were also specified.

An effective starting point for the lubrication program is an inventory of the equipment in the plant. Classify or group together items in the same department or functional requirement.

Obviously, first priority is given to the most important equipment. Major production equipment would probably have priority in most plants. Environmental control items would also be high on the list. For instance, you might begin with the central power plant and group together auxiliary equipment for the steam generation system and the equipment generating chilled water.

Next, record all available information that relates to lubrication of the equipment items listed. Information sources might include master equipment record cards, manufacturers' operating manuals, and the present lubrication procedure for each item. Usually this inventory will uncover many items that have little, if any, recorded information.

There is a high degree of probability that the lubrication procedures for unrecorded equipment are wasteful, inefficient, or nonexistent and are leading to early breakdown or retirement of the equipment.

Before proceeding further into the specifics of a lubrication program such as establishing proper lube cycles, evaluating manpower requirements and operating the entire program, become familiar with some of the basic lubricant definitions, basic theory, and practical tables/charts, to aid in understanding this important subject.

TYPES OF LUBRICANTS

Lubricants can be broken down into four generic types: gaseous, liquid, plastic, and solids.

Gaseous lubricants. These will not be given consideration here because they are used in very restricted applications such as very light bearings, extremely high speed spindles, inertial gyroscopes, etc.

Liquid lubricants. The major group of liquid lubricants are made up of the mineral oils produced from petroleum crudes by various refining processes. These crudes are complex compounds of carbon and hydrogen, referred to as hydrocarbons. A secondary group of liquid lubricants are called fixed oils. These are fatty substances extracted from animals, vegetable matter, and fish. They are called fixed oils because they will not volatilize without decomposing. This chapter concentrates on nondrying fixed oils, which are usually blended with mineral oils for specific requirements. A third and rapidly growing classification of liquid lubricants is synthetic fluids. These fluids have been chemically developed to meet the tremendously growing demands placed on lubricants caused by technology developments in equipment of all types, such as increased speeds, pressure, and temperatures. In some applications, the use of synthetic lubricants is mandatory. For example, mineral petroleum oils cannot be used in lubricant systems exposed to ignitable environments, extreme heat, or extreme cold.

Synthetic oils must be used where high temperature may carbonize mineral oils. Such a breakdown will cause maintenance problems or failure in equipment such as ovens, draft dampers, ammonia and air compressors, or other high-temperature-oriented equipment.

Plastic Lubricants. This is a general classification for greases, petrolatum and semi-fluids. The primary interest for this chapter is in greases. Plastic lubricants are indicated where operating conditions make it difficult to keep a lubricant in place: for example, in a bearing located in such a position that any oil would leak right out.

Greases are generally made by saponifying* a metal base with a fatty substance to form a soap to which any of the liquid lubricants discussed above, are added. This process, which includes agitation, is performed under controlled temperature conditions either in a pressure cooker or an open kettle. Some soaps are formulated to make greases withstand high temperatures; others give a waterproof (or steamproof) characteristic to the finished grease. A variety of application possibilities can be designed into greases by varying base oils and/or saponifying

---

*A glossary of terms pertinent to lubrication appears at the end of this chapter.

agents. In most instances, the grease selected should be recommended by the grease manufacturer under the guidance of the equipment manufacturer.

Solid Lubricants. These include such products as graphite, molybdenum disulfide, polytetrafluorotheylene (PTFE) talc, mica, borax, and wax. Solid lubricants are sometimes used in their dry states such as molybdenum disulfide, graphite, PTFE, but more often they are used as additives in oils and greases or are combined with a vehicle to place them on a specific surface and keep them there. Dry solid lubricants should be applied on surfaces subjected to high temperatures (such as oven chains) or in dusty environments where an oil or grease would attract dirt and hold it there.

VISCOSITY AND VISCOSITY INDEX

The measurement of viscosity and viscosity index is fundamental to the science of lubrication. Viscosity is the property of a fluid that resists the force, tending to cause the fluid to flow. It is also a measurement of the rate at which a fluid is capable of flowing under specified conditions. Fluidity decreases with higher viscosity; for example, flow rate varies inversely with viscosity, and vice versa.

Viscosity and viscosity indices are very important to the lubrication engineer. Viscosity plays a major role in developing the film thickness of a hydrodynamic wedge. It determines load-carrying capacity, the ability of an oil to flow, operating temperature levels, and often wear rate. Generally, high viscosity or heavy oils are used on parts moving at slow speeds under high loads, since the heavy oil resists being squeezed out from between the rubbing parts. A high viscosity index is required when a wide range of operating temperatures is indicated. Low viscosity or light oils are used when higher speeds and lower loads are encountered; they do not impose as much drag on high speed parts and the high speed permits a good oil wedge to form, even though the oil is less viscous.

Low temperatures call for lighter oils and high temperatures call for heavier oils. Oils increase in viscosity as their temperatures become lower and decrease in viscosity as their temperatures go up.

Kinematic viscosity is the time required for a fixed amount of an oil to flow through a capillary tube under the force of gravity. The unit of kinematic viscosity is the stoke or centistoke (1/100 of a stoke). Kinematic viscosity may be defined as the quotient of the absolute viscosity in centipoises divided by the specific gravity of a fluid while both are at the same temperature.

As stated above, viscosity is a measure of the "flowability" at a definite temperature. The unit of measure is time in seconds required for 60 milliliters of oil to flow through a standard orifice under a standard falling head and at a given temperature. (100°F and 210°F are common temperatures for reporting viscosity.) Saybolt Universal viscosimeters heat a test sample to the desired temperature and then pass the sample through the orifice while being timed. Saybolt Furol viscosity is obtained with the same instrument but with a larger orifice, producing results approximately 1/10 those of the Universal orifice readings.

SAE	SUS @ 100°F	SUS @ 210°F	CPS (Centipoise)*
10	200	48	44
20	325	57	70
30	550	68	119
40	850	84	184
50	1200	100	260
90	1500	110	326
140	2500	150	543
250	5000	200	1085

*Equivalent to SUS @ 100°F

Figure 9-1.

Viscosity Conversion Chart.

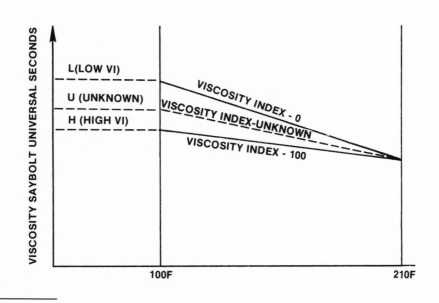

*From Lubrication by Gunther

Figure 9-2.

Viscosity Index.

Lubricating oils decrease in viscosity as their temperatures are raised. For example, an oil having a viscosity of 3,000 seconds at 100°F (SUS) might show a viscosity of 175 sec. at 210°F. This is said to be the rate of viscosity change for that oil. Another oil of 3,000 seconds at 100°F may have a viscosity of 160 seconds (SUS) at 210°F. This oil then would be considered to have a higher rate of viscosity change per degree temperature rise.

The rate at which an oil changes viscosity with a rise or drop in temperature is designated by a comparative number called viscosity index (VI). A VI of 100 indicates that an oil with this value thins out less rapidly than oils having a VI of zero. Values in between indicated intermediate viscosity change according to the VI number. (See Figure 9-2.)

Figure 9-3 shows viscosity versus temperature relationship for SAE (Society of Automotive Engineers) multigrade motor oils. Note an SAE 30 oil compared to an SAE 10W40 motor oil. The 10W40 oil has more body (higher viscosity) at 210°F to

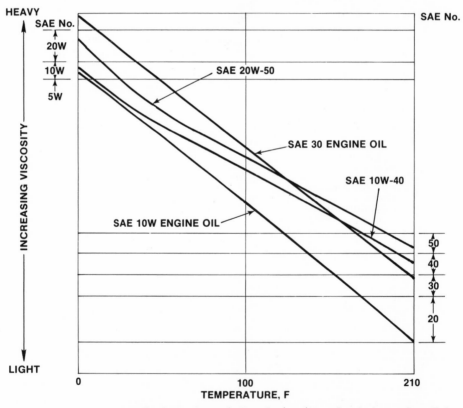

From Classification Systems for Auto, Truck, and Heavy Duty Equipment, Power Train Lubricants by Sun Petroleum Products Company, A Division of Sun Oil Company of Pennsylvania.

Figure 9-3.

Viscosity—Temperature Relationship.

provide more protection at high temperatures and a lower viscosity at 0°F to provide faster starts, and less wear in the winter cold because of better "pumpability." An SAE 10W40 oil has a higher VI than an SAE 30 oil.

MECHANISM PARTS REQUIRING LUBRICATION

All surfaces in any mechanism that move relative to other surfaces require lubrication. This is to separate these surfaces with a film of lubricant to reduce friction and wear. This occurs in bearings that support rotating shafts, in gears that have meshing teeth, and between pistons and the cylinders in which they move. All mechanisms, regardless of size or complexity, will employ one or more of the following basic moving parts: bearings, gears, and pistons.

I.   Bearings:

These can be divided into two basic types, plain bearings and anti-friction bearings.

A. Plain bearings can be further classified according to function. Journal, guide, and thrust bearings are usually made from bronze, babbit, plastic (nylon, teflon), or some material softer than steel so that any wear will be on the bearing instead of the steel part.

Journal bearings are so named because they support or operate against a rotating shaft. The portion of the shaft within the bearing is known as the journal. In a machine, when a journal starts to rotate, a wedge-shaped film of oil develops under the journal. This wedge lifts the journal away from the bearing, reducing friction, and thus guarding the journal and bearing against wear. Journal bearings are of various designs:

1.   The *solid bearing* as the name implies, is in one piece, and is sometimes called a *sleeve* bearing or a *bushing*.

2.   The *split bearing* is divided into two pieces lengthwise.

3.   The *half bearing* encircles only one-half of the journal, leaving the other half-exposed. It is used when the load is applied vertically between the bearing and the shaft.

4.   The *multipart bearing* normally consists of four separate parts or quarters. They are found on crankshafts on such machines as air compressors and steam engines. The four-piece construction allows for readjustment of the bearing to the crankpin to take up slack when wear takes place.

Guide bearings guide or hold in proper position reciprocating parts of a machine.

Thrust bearings are bearings that prevent a shaft from moving endwise.

B. Anti-friction bearings are bearings in which a series of balls or rollers are interposed between the moving parts. These balls or rollers are usually (but not always) mounted in a cage and enclosed between rings known as races. Normally the outer race is fixed in the equipment while the inner race is tightly fixed on the shaft with which it rotates. When the shaft rotates, the balls or rollers are rotating between the inner and outer races with a minimum of friction. Antifriction bearings are classified as follows.

1. *Straight roller bearings* are where the axis of the rollers are parallel to the axis of the bearing.

2. *Tapered roller bearings* are used not only to support a rotating shaft but also to prevent the shaft from moving endwise.

3. *Ball bearings* are probably the most commonly known and used of all antifriction bearings. They come in single- and double-row and in self-aligning types.

4. *Ball thrust bearings* are usually flat bearings where the loads are applied to the flat of the race rather than the inner and outer diameters, and as the name suggests, are designed to prevent linear movement of the shaft rather than for radial movement.

5. *Needle bearings* differ from the others in that they have no inner race, cage, or separator, and their small rollers or needles are just slightly separated by the lubricant. The name comes from the fact that the rollers are much longer than their diameter. They are primarily used for oscillating and intermittent motion.

II. Gears:

Gears serve various functions, such as transmission of motion from one shaft to another; change of direction and orientation from one shaft to another; or change of speed from one shaft to another. Properly matched gear tooth surfaces are designed to provide a rolling motion for minimum tooth friction.

III. Pistons:

The third fundamental moving part found in machinery is the piston, which operates in a cylinder. In an automotive engine, for example, fuel and air mixtures are pumped in above the piston and ignited. Combustion forces the piston to move downward to turn a crankshaft. Rings are located in grooves in top of the piston pushing outward against an oil film to seal the space between the cylinder wall and the piston and prevent loss of power escaping downward. The lubricating film of oil used to separate the rings from the wall of the cylinder along which it must slide also helps the rings seal the combustion chamber described above. The oil may be splashed onto the rubbing surfaces from the crankcase; however, in large pistons it is often pumped in small quantities through holes in the piston wall. Similar piston action may be found in other applications such as compressors, pumps, and even in a grease gun.

LUBRICATION REQUIREMENTS FOR MECHANISM PARTS

The lubricating instructions of the machine manufacturer should always be sought and followed. Lubricant suppliers can be helpful with suggestions for equivalencies, and sometimes even improvements through new lubrication products.

Grease Lubrication of Anti-Friction Bearings

There are several rules associated with grease lubrication of anti-friction bearings:

1. Use the correct grease according to the type of rolling element and the service for which it is used. Normally a grease should contain an oil having the same viscosity that would be used if the oil alone were providing the lubrication. Never use a grease containing solids in any needle bearing or in a roller bearing turning more than 100-150 rpm.

2. Use a grease of the lowest possible consistency commensurate with the temperature, speed, leakage rate, and position of the bearing. Lower consistency greases generally contain more oil, and oil is the principal lubricant upon which bearing life depends.

3. Never fill an anti-friction bearing to full capacity. Otherwise, it will run hot and subsequent expansion of the grease will eventually break the seals.

   The proper volume of grease in a bearing should be approximately equal to one-third to one-half of its space capacity. The exception to this are low-speed bearings where the cavity may be nearly filled to capacity.

   There is no one simple answer to knowing when to stop filling an anti-friction bearing except perhaps in a newly hand-packed bearing. One way of assuring against overpacking a bearing is by providing a pressure relief opening at the bottom of the housing; a threaded plug is temporarily removed from this opening during the greasing process. After greasing, the relief opening is left open for a short time to allow any excess grease to be forced out. The flow of grease will stop when the pressure in the housing has found its own comfortable level. The plug should then be replaced in the relief opening. Another method involves the use of an automatic lubing device, several types of which are on the market. These devices replace grease fittings and have their own reservoir and metering devices to feed grease to the bearing. Some have springs, some have pressure equalizing chemicals, and others have a diaphragm into which the grease is pumped. When inadequate grease is in the bearing, it creates an imbalance in pressure, causing the atmospheric pressure on the diaphragm to push the right amount of grease into the bearing to reestablish and maintain equilibrium.

4. Depending on the presence of water and other contaminants in the environment of the bearing during operation, the greasing interval should be planned to allow one complete change of grease scheduled on a basis of shaft rotation. In an extremely dirty environment this could mean full purging at every application, whereas purging at every eighth application might be sufficient in a clean, dry environment. The purging can be performed as described above, i.e, feeding the grease into the bearing until all the old grease is removed or the new grease appears at the drain.

5. Before attaching a grease gun to a grease fitting, wipe the nozzle and the fitting clean with a lintless cloth, preferably dampened with an industrial cleaning solvent. This is a precaution against pumping foreign matter into the bearing along with the grease.

Lubrication of Gears

The American Gear Manufacturers Association (AGMA) has published the following standards for enclosed gear drivers: *AGMA Standard Specification— Lubrication of Industrial Enclosed Gear Drives.* The following material has been extracted from AGMA Spec (AGMA 250.03) with the permission of the publisher, The American Gear Manufacturing Association, 1330 Massachusetts Ave., N.W., Washington, D.C. 20005. This standard covers the lubrication of industrial enclosed gear drives having the following types of gearing:

- Helical
- Herringbone
- Straight Bevel
- Worm
- Spur

These lubrication recommendations apply only to enclosed gear drives that are designed and rated in accordance with current AGMA standards. Gear drives operating at speeds above 3,600 rpm and/or pitch-line velocities above 5,000 fpm (feet per minute) may require additional consideration.

The ambient temperature range is −40°F to +125°F, and is defined as the air temperature in the immediate vicinity of the gear drive. Gear drives exposed to the direct rays of the sun will run hotter and must therefore be given special consideration. (Maximum lubricant sump temperatures are defined in Figure 9-4.)

Gear drives operating outside of these temperature ranges, or those operating in extremely humid, chemical, or dust-laden atmospheres should be referred to the gear drive manufacturer.

Lubricant viscosity recommendations are specified by AGMA Lubricant Numbers. The corresponding viscosity ranges are shown in Figure 9-5. AGMA Lubricant Number recommendations for drives using all types of gearing except worm gearing are given in Figure 9-6 (for R & O gear oils and EP gear lubricants).

Lubricant	Maximum Sump Temperature °F
Rust and oxidation inhibited (R & O) gear oil	+200
EP gear lubricants	+160 to 200*
Compounded gear lubricants	+160 to 200

*Maximum allowable sump temperature depends upon the particular lubricant selected. Sulphur-phosphorus type EP (extreme pressure) oils are generally more stable than lead-naphthenate type EP oils or compounded gear oils.

Extracted from AGMA Specifications—Lubrication of Industrial Enclosed Gear Drives (AGMA250.03), with permission of the publisher, The American Gear Manufacturers Association, 1330 Massachusetts Avenue, N.W., Washington, D.C. 20005.

Figure 9-4.

Maximum Lubricant Sump Temperatures.

Rust and Oxidation Inhibited Gear Oils	Viscosity Range ASTM System (2)	Extreme Pressure Gear Lubricants (3)
AGMA Lubricant No.	SSU at 100°F	AGMA Lubricant No.
1	193 to 235	
2	284 to 347	2 EP
3	417 to 510	3 EP
4	626 to 765	4 EP
5	918 to 1122	5 EP
6	1335 to 1632	6 EP
7 comp. (1)	1919 to 2346	7 EP
8 comp. (1)	2837 to 3467	8 EP
8A comp. (1)	4171 to 5098	

(1) Oils marked comp. are compounded with 3% to 10% fatty or synthetic fatty oils.

(2) "Viscosity System for Industrial Fluid Lubricants", ASTM 2422. Also British Standards Institute, B.S. 4231. Viscosity ranges for AGMA lubricant numbers will henceforth be identical to those of the ASTM System.

(3) Extreme pressure lubricants should be used **only** when recommended by the gear drive manufacturer.

Extracted from AGMA Specifications—Lubrication of Industrial Enclosed Gear Drives (AGMA250.03), with permission of the publisher, The American Gear Manufacturers Association, 1330 Massachusetts Avenue, N.W. Washington, D.C. 20005.

Figure 9-5.

Viscosity Ranges for AGMA Lubricants.

Type of Unit Size of Unit	Other Lubricants		AGMA Lubricant Number	
			Ambient Temperature °F	
Main Gear Low Speed Centers	-40 to 0	-20 to +25	15 to 60	50 to 125
Parallel shaft, (single reduction), up to 8 in.			2–3	3–4
Over 8 in. and up to 20 in.,			2–3	4–5
Over 20 in.			3–4	4–5
Parallel shaft, (double reduction), up to 8 in.	Automatic Transmission Fluid (or similar product—see note 5)	SAE 10W/30 or 10W/40 Motor Oil (or similar product—see note 5)	2–3	3–4
Over 8 in. and up to 20 in.			3–4	4–5
Over 20 in.			3–4	4–5
Parallel shaft, (triple reduction), up to 8 in.			2–3	3–4
Over 8 in. and up to 20 in.			3–4	4–5
Over 20 in.			4–5	5–6
Planetary gear units				
O.D. Housing up to 16 in.			2–3	3–4
O.D. Housing over 16 in.			3–4	4–5
Spiral or straight bevel gear units				
Cone distance up to 12 in.			2–3	4–5
Cone distance over 12 in.			3–4	5–6
Gearmotors and shaft mounted units			2–3	4–5

NOTES

1. Pour point of lubricant selected should be at least 10°F lower than the expected minimum ambient starting temperature. If ambient starting temperature approaches lubricant pour point, oil sump heaters may be required to facilitate starting and insure proper lubrication.
2. Ranges are provided to allow for variations in operating conditions such as surface finish, temperature rise, loading, speed, etc.
3. AGMA viscosity number recommendations listed above refer to R&O gear oils. EP gear lubricants in the corresponding viscosity grades may be substituted where deemed necessary by the gear drive manufacturer.
4. When they are available, good quality industrial oils having similar properties are preferred over the automotive oils. The recommendation of automotive oils for use at ambient temperatures below +15°F is intended only as a guide pending widespread development of satisfactory low temperature industrial oils. Consult gear manufacturer before proceeding.
5. Drives incorporating overrunning clutches as backstopping devices should be referred to the clutch manufacturer as certain types of lubricants may adversely affect clutch performance.

Extracted from AGMA Specifications—Lubrication of Industrial Enclosed Gear Drives (AGMA250.03), with permission of the publisher, the American Gear Manufacturers Association, 1330 Massachusetts Avenue, N.W. Washington, D.C. 20005.

Figure 9-6.

AGMA Lubricant Number Recommendations for Enclosed Helical, Herringbone, Straight Bevel, Spiral Bevel and Spur Gear Drives.

Extreme Pressure Gear Lubricants: Extreme pressure gear lubricants are petroleum-based lubricants containing special chemical additives. EP gear lubricants recommended for enclosed gear drives are those containing either lead naphthenate of sulfur-phosphorous additives. EP gear lubricants should be used only when specified by the gear drive manufacturer.

Compounded Gear Oils: Compounded gear oils are a blend of a petroleum-based lubricant with 3 to 10 percent fatty or synthetic fatty oils. These lubricants are usually used for worm gear drives.

Synthetic Gear Lubricants: Diesters, polyglycols and synthetic hydrocarbons have been used successfully in enclosed gear drives for special operating conditions.

Lubricant Maintenance

Lubricant change intervals. The lubricant in a new gear drive should be drained after four weeks' operation. The gear case should be thoroughly cleaned with a flushing oil. The original lubricant can be used for refilling if it has been filtered through a filter of 100 microns or less (50 microns or less for high speed gear units using sleeve bearings); otherwise, new lubricant must be used. Lubricants should not be filtered through Fullers Earth or other types of filters that remove lubricant additives.

Under normal operating conditions, the lubricant should be changed every 2,500 hours of operation or every six months, whichever comes first. Extended change periods may be established through periodic testing of oils.

A rapid rise and fall in temperature may produce condensation, resulting in the formation of sludge. Dust, dirt, and chemical fumes also react with the lubricant. Sump temperatures in excess of those listed in Figure 9-4 will result in accelerated degradation of the lubricant. Under these conditions, the lubricant should be changed every one to three months, depending on severity.

Cleaning and flushing. The lubricant should be drained while the gear drive is at operating temperature. The drive should be cleaned with a flushing oil.

Used lubricant and flushing oil should be removed completely from the system to avoid contaminating the new charge.

Avoid the use of a solvent unless the gear drive contains deposits of oxidized or contaminated lubricant that cannot be removed with a flushing oil. When persistent deposits necessitate the use of a solvent, a flushing oil should then be used to remove all traces of solvent from the system.

The interior surface should be inspected where possible, and all traces of foreign material removed. The new charge of lubricant should be added and circulated to coat all internal parts.

Cold Temperature Starting

Low temperature gear oils. Gear drives operating in cold areas must be provided with oil that circulates freely and does not cause high starting torques. An acceptable low-temperature gear oil, in addition to meeting AGMA specifications, must have a low pour point and a viscosity low enough to allow the oil to flow freely at the startup temperature but high enough to carry the load at the operating temperature.

Sump heaters. If a suitable low-temperature gear oil is not available, the gear drive must be provided with a sump heater to bring the oil up to a temperature at which it will circulate freely for starting. The heater should be so designed as to avoid excessive localized heating, which could result in rapid degradation of the lubricant.

TYPES OF EQUIPMENT AND LUBRICANTS TO BE USED

Various pieces of equipment likely to be found in a typical operation should be considered. The parts to be lubricated and the lubricants to use are as follows:

Electric Motors

Proper lubrication is most important because most motors employ precision-built, anti-friction bearings and contain insulation on the windings, both of which can be damaged. Therefore, in order to keep the electric motor operating properly, the bearing must be lubricated properly. The lubricant must protect the bearing and, therefore, must stay in the bearing. If it leaks out onto the windings, it can interfere with cooling air flow, causing overheating and possible short circuiting or fire.

It is important that motor bearing housings should be sealed against dust, dirt, and grit. Because of this, the sealed-type, prelubricated ball bearing is frequently used, particularly in adverse environments, where speed control and low power consumption are important. Recommendations are typical but will vary for some applications according to temperature and speed range.

Bearings—Oil Lubricated (speeds up to 3,600 rpm)

Ambient Temperature (-°F)	Viscosity (SUS at 100°F)
Below 32	150 to 225 (pour -15°F)
32–200	300

Grease Lubricated (speeds up to 1800 rpm)

Ambient Temperature (-°F)	Consistency (N.L.G.I. No.)
-10 to 32	0
32 to 200	2

Air Compressors

Compressors are divided into three basic types: reciprocating, screw, and rotary vane. In all instances, the oxidizing effect of air and high discharge temperatures tend to cause deterioration of the cylinder lubricating oil. Overapplication promotes the formation of carbonaceous deposits in the cylinder head, piston rings, and winds up, as well, out on the discharge valves, in the valve pockets, and on the valve springs.

Air compressor cylinder lubrication requires an oil of high chemical stability, high viscosity index, low carbon content, high flash point, and high thermal stability. To achieve the foregoing, a highly refined petroleum or synthetic oil containing corrosion and rust inhibitors is required.

Reciprocating compressors handling dry air, inert gases, nitrogen, CO_2, etc., (not oxygen), up to 150 psi discharge pressure may be lubricated adequately with a nondetergent oil having a viscosity of 300 SUS at 100°F. For medium range pressures (up to 2,000 psi) use an oil of 600 SUS at 100°F; between 2,000 and 7,000 psi (cryogenic ranges) a 1,200 to 1,500 SUS at 100°F is required. Lubrication of crankcase components of low-pressure compressors (up to 150 psi) use the same oil as the cylinders. For higher pressure units, use a 600 SUS at 100°F oil regardless of application.

Screw compressors usually require a lower viscosity lubricant than a reciprocating compressor. A 150-200 SUS at 100°F oil is suggested here.

Rotary compressors require lubrication of both bearings and rotor cylinder. For most services, including refrigeration, a 300 SUS at 100°F nondetergent oil is recommended.

Refrigeration Compressors

Proper lubrication for the moving parts of the refrigeration compressor is affected by a number of problems that relate to the thermal process of the refrigeration system. The piping and equipment beyond the compressor are involved, in addition to the internals of the compressor itself. Many problems are caused by improper oil selection. The interreaction of some refrigerants with oils, and/or incorrect piping designs may cause a number of problems, such as:

a. *Loss of volumetric efficiency.* Vaporized oil mixed with the refrigerant reduces the rate of flow of refrigerant handled by the compressor.

b. *Heat transfer losses.* A film of oil, or any of its constituents, including wax deposited on heat transfer surfaces, reduces the heat transfer efficiency of evaporators, condensers, and other heat exchangers.

c. *Deterioration of dryers.* Desiccants lose their ability to absorb moisture when pores become oil-clogged.

d. *Expansion valve clogging.* Poor quality oils containing waxes can cause trouble when exposed to solvent-type refrigerants. Wax deposits on expansion valves and other internal moving parts will cause them to malfunction.

e. *Line clogging.* When refrigerant flows too slowly oil drops out in the refrigerant lines and drains into the lower parts of the system, interfering with refrigerant flow.

f. *Foaming.* Foaming increases oil carryover to the discharge side of the system and reduces the lubricating quality of the oil.

g. *Carbon deposits.* Carbonaceous deposits form on discharge valves, cylinder heads, pistons and in discharge lines, and are caused by oil deterioration under high temperature. This condition can cause serious damage.

h. *Cylinder blow by.* Loss of compression results from the escape of gas around the piston back to the crankcase. This can be caused by insufficient or improper lubricant. In consideration of the above, oils selected for refrigeration compressor applications must possess the following properties:

Chemical stability	High dielectric strength
Low pour point	Proper viscosity
Low floc point	Low volatility

Air Handling Units

Fans

Propeller and disk fans are sometimes mounted directly on the motor shaft, in which case only the electric motor bearings need be considered for lubrication. The motors of small fans are often factory lubricated for the life of the rolling bearings. An oil of 150 to 300 SUS at 100°F is recommended for oil-lubricated bearings applied by inverted wick, ring, or flooded periodically by oil can. A no. 2 NLGI lithium or calcium complex grease is recommended for temperatures between 40° and 200°F and speeds ranging from 600 to 3,600 rpm. For lower temperatures, use a No. 1 NLGI grease of similar soap bases and a high temperature grease for temperatures above 220°F.

Centrifugal fans, used primarily for ventilation, are considered light-duty types and are mounted on their own bearings which, are usually ring oil sleeve type. A 300 SUS at 100°F oil is recommended for ambient temperatures of 40° to 100°F, a 600 SUS at 100°F oil is recommended for ambient temperatures above 100°F. Centrifugal fans of the heavy-duty types, including forced and induced draft fans and large evaporative condenser fans, may be equipped with either sleeve or rolling bearings.

Basic lubrication requirements are as follows:

Bearing Type and Condition	Oil Viscosity (SUS) at 100°F
Sleeve bearings, ring-oiled, water-cooled temperatures 32-90°F	300–400
100 to 150°F ambient	400–600

Bearing Type and Condition	Oil Viscosity (SUS) at 100°F
Sleeve bearings, ring-oiled, not water-cooled*	600
Sleeve bearings, oil circulation with oil cooler	200–300
Roller bearings** oil bath:	
40 to 140°F ambient	300
140 to 220°F ambient	900

*Fans handling hot gases up to 150°F, such as induced-draft fans, may require an oil from 900 to 1,500 SUS at 100°F. This type of application invites mineral oil deterioration because of oxidation. Specially formulated synthetic oils should be considered for this type of duty.

**Fans handling hot gases or subject to temperature above 150°F require a heavier oil of 1,500 to 1,800 SUS at 100°F, preferably synthetic.

Blowers

The oil viscosity for blower bearings and geared transmissions depends on the horsepower rating of the blower. For 25 horsepower or less, use a 600 SUS at 100°F oil; for over 25 horsepower use a 900 SUS at 100°F oil. Turbo-blowers, single- or multi-stage, are constant pressure machine driven by an electric motor or steam turbine. Use a circulating oil of 150 SUS at 100°F for bearing speeds of 1,800 rpm and up. For speeds lower than 1,800 rpm, ring-oiled bearings use an oil of 300 to 400 SUS at 100°F.

Pumps

Liquid pumps of the horizontal reciprocating types require lubrication as follows:

Crossband, crank, gears, and bearings	300 SUS 100°F
Vertical centrifugal pumps	
upper bearing	300 SUS 100°F
lower bearing	Grease #2 lime or calcium complex
Horizontal centrifugal pumps	Grease #2 calcium complex

Couplings

Rigid and flexible couplings usually do not require lubrication. Flexible couplings with articulated joints attain flexibility by mechanical means and do require lubrication. Factors of temperature, angular velocity load transmitted, radial positions, and location of frictional parts, effectiveness of seals, etc., will affect the selection of a lubricant. The use of a lubricant specified by the coupling manufacturer is recommended. If the application involves unusual operating conditions (heavy shock loads, frequent axial movement, large speed variations, or extreme temperatures), such data should be submitted to the manufacturer when requesting lubricant recommendations.

The following are general suggestions for coupling lubrication. Cold operation in low temperature climates, cold storage rooms, high altitudes, etc., requires an oil with a pour point below 0°F and viscosity less than 220 SUS at 100°F. Intermediate temperature operation between 0 and 40°F requires a 225 to 250 SUS at 100°F lubricant. Low pour point leaded compounds of appropriate viscosity are also recommended. Temperatures between 40 and 120°F require an oil of 600 to 2,500 SUS at 100°F or a grease. A lead lime grease No. 0 or No. 1 NLGI consistency or a No. 2 aluminum soap-base grease is preferred. A leaded compound having a viscosity of about 400 SUS at 210°F is also highly satisfactory.

Chain Drives

Chain drives should be lubricated such that the rollers are free-turning at all times. When the roller reaches the sprocket, it should not turn relative to the sprocket; all motion should be inside the roller. Frequently, a grease or oil applied to a chain will attract and hold airborne foreign matter that will gum up, so that these rollers do not turn freely. This causes excessive wear to the sprockets and the chain. A penetrating-type oil with a dry lubricant such as molybdenum disulfide, graphite, or PTFE suspended in it will lubricate the chain internally, thus keeping the rollers free-turning without attracting foreign matter to the chain. The vehicle used should contain rust and oxidation inhibitors to leave a protective coating on all metal surfaces.

Hydraulic Systems

The majority of industrial hydraulic systems, including elevators, perform satisfactorily with oils having viscosities in the range of 150 to 325 SUS at 100°F for operating temperatures up to 165°F. In most cases a 10W40 grade motor oil (API service classification SE/CC) will operate in the wide temperature range that elevator hydraulic systems require for all seasons. Follow the equipment manufacturer's recommendations.

Other systems require the following:

Vane pumps	150-300 SUS at 100°F
Angle and radial piston pumps	150-900 SUS at 100°F
Axial piston pumps	150-300 SUS at 100°F
High pressure, high output pumps	300 SUS at 100°F
Gear pumps	300-600 SUS at 100°F

Pneumatic Tools

Oils that are especially compounded to lubricate pneumatic tools and machine power cylinders are called air-line oils. They are similar to turbine-type oils, containing a small amount of emulsifying agent to combat the washing effect of condensate that precipitates from compressed air upon expansion. Viscosity ranges from about 200 SUS at 100°F for small tools to about 800 SUS at 100°F for larger cylinder wall areas.

Laundries

Laundries contain various types of equipment that require special lubrication. Washing machine and water extractors, where lubricated, should use a waterproof grease, preferably made with an aluminum complex soap base. Mangles and/or large industrial ironing machines require lubrication frequently. They have rolls at each end with pillow block bearings over which steam pours. These bearings should also use a waterproof grease with good heat resistance.

Kitchens

Here the primary concern is to use U.S.D.A. AA approved lubricants where possible contact with food might occur. Commercial kitchens, such as in hospitals or schools, frequently have conveyors that have the usual complement of motors, gear boxes, bearings, plus chains of various designs and configurations. Care should be taken to use waterproof lubricants in areas where steam or water exist, such as commercial dishwashers.

PLANNING THE PREVENTATIVE MAINTENANCE PROGRAM

The plant maintenance supervisor must have an accurate inventory and record of all plant equipment.

All equipment to be lubricated should be listed. Equipment location should be identified by department and/or function: i.e., power house, machine shop, material handling equipment, HVAC, etc. Work center numbers can be assigned to these areas. Numbers can be assigned to the individual pieces of equipment within a given work center. Depending on the number of work centers and machines, these can be combined to make one number where the first or first two digits are the work center and the last digits are the individual piece of equipment in that work center. If convenient, a dash can be placed between the work center number and the machine number to improve legibility.

Starting with the power source (usually a motor) and working outward, each component of the equipment can be listed and the parts to be lubricated noted. It is desirable, but not always necessary, to indicate the number of points, such as oil holes, grease fittings, pillow blocks, and couplings. The correct lubricant to be used and the preliminary cycle for lubrication should also be noted. Provision should be made to record the results of oil analysis by a qualified laboratory.

Schedule a period of time in which to evaluate and update the survey. Card systems (see Figure 9-7) or computer printouts should be implemented so that the person performing the actual lubrication can periodically be handed a series of cards or a printout sheet listing his performance requirements for that day/week. With a card system, there can be a form on the back of the instruction card for date and initialing to verify the actual lubrication or inspection performed.

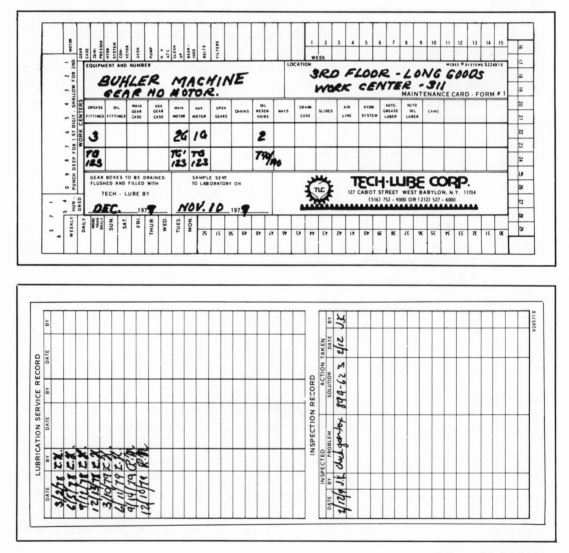

Courtesy of Tech-Lube Corp.

Figure 9-7.

Keysort-Lubrication Card.

Space could be provided on the card to note service performed because of failures or overheating or during periodic servicing of the machine or component. This becomes a handy log on that component for trend analysis and program update.

Once the effectiveness of the program has been established, the lubricants in the storeroom, the transfer equipment to transport the lubricant from the storeroom to the machinery, and the actual lubrication point on the equipment should be color-coded to simplify the oiler's job and to reduce the possibility of using an incorrect lubricant.

TRAINING MAINTENANCE PERSONNEL IN LUBRICATION PROCEDURES

Mechanics should be aware of techniques of lubricant application; operating principles and maintenance of application devices; the importance of maintaining proper oil levels and lubricant feed rates; and the limitation of grease application to rolling bearings. They should also have a basic knowledge of lubricants, and the uses and necessity of applying each lubricant to its proper frictional component.

An initial period of on-the-job training with an experienced mechanic, followed by attendance at outside clinics or seminars, is an effective method of training new lubricating mechanics. Clinics and seminars are held by representatives of lubricant suppliers and equipment manufacturers, such as Tech-Lube Corp., West Babylon, New York and Imperial Oil and Grease Company, a Beatrice Chemical Company/Division of Beatrice Foods Company, Los Angeles, California, and others. Such clinics are usually divided into a number of weekly sessions of approximately one hour each to provide a progressive training program and can be augmented by associated reading matter.

SCHEDULING

Lubricants are consumed or deteriorate from use. Consumption is caused by leakage (which can also have its own serious harmful side effects), evaporation, blow by and burning. Deterioration is caused primarily by heat, which with the catalysts of water and air, promotes oxidation and forms acids.

All-loss lubrication systems and defective piping, joints, and seals of circulation systems represent the areas of highest lubricant consumption. Be sure equipment is properly vented and sufficiently cooled. In the case of all-loss oil and grease application systems, the rate of lubricant consumption under normal operating conditions is fairly regular. On the basis of probability, such equipment will require lubrication at predictable intervals, whether in terms of hours, days, or months, depending on the equipment and its application.

Similarly, the life of a lubricant in a *reuse system* is a function of operating conditions. Where such conditions remain relatively constant, the lubricant cycles of that system are predictable within safe limits. Again, it requires experience, guided by periodic laboratory analysis, to determine the probable life span of a given system.

The time between introduction of an oil to a given system and the point when it reaches a specific level of deterioration, as determined by laboratory analysis, establishes the frequency of drain periods for that system and for that particular oil. Oil change can be scheduled correctly only after the interval is determined by competent chemical analysis. Ball and roller bearings of electric motors and machine spindles are often overlubricated. The trend, as previously noted, is to prepack and adequately seal such bearings, but they should be evaluated periodically by monitoring their operating temperatures. Mechanics should become familiar with all the lubricants used and where each is applied. Such information is found on the lubrication charts drawn up as a result of the plant-wide analysis previously mentioned. These charts should be available to the oiler mechanic and updated periodically by his findings and recommendations.

SOME COMMON MISTAKES IN LUBRICATION PROGRAMS

Use of the wrong lubricant—Do not use a common motor oil in a compressor; a high-temperature grease in a wet environment (such as a water pump); or a light-weight oil in a heavy-duty, slow-moving gear box. Care must be taken to follow proper directions.

Improper lube frequencies—The lube frequencies previously suggested are for guidance only. The final schedule must be determined from actual operational experience and chemical analysis. Adhere to the final schedule as rigidly as possible.

Overlubing and underlubing—It is very important to understand that overlubrication can be as bad and sometimes even worse than underlubrication. If a certain amount of lubrication is good, it doesn't necessarily follow that more is better.

Proper inspection—The system oil levels in gear boxes should be checked regularly, bearings should be checked for overheating, and lines should be checked for breakage. It is important that the program be reviewed and monitored by a qualified individual in order to ensure success.

Mixing lubricants—It is inadvisable to mix lubricants. Lubricants should be checked thoroughly before mixing to confirm compatibility, viscosity, performance, etc. Lubricant suppliers can frequently be helpful in this connection, but an independent chemical oil analysis should be performed when in doubt.

Looking at price, not cost—Real cost cannot be measured by price alone. Bottom-line cost at the end of the year is the criterion by which the purchase price of a lubricant should be evaluated.

SUMMARY

A PM program using adequate inspection, proper lubricants in the correct place, and quantities at the right time can save the organization many dollars over a time span. It can make the oiler's job more routine and reduce the possibilities of human error. This, with better performance, will reduce energy consumption, downtime, replacement parts cost, and prolong the smooth operation of the equipment.

GLOSSARY OF LUBRICATION TERMINOLOGY

This glossary covers some of the more important terms in lubrication.

Bleeding: The tendency of a liquid component to separate from a liquid solid or liquid semisolid mixture, as an oil from a grease.

Boundary lubrication. A condition of lubrication whereby the friction between two relatively moving surfaces is determined by the properties of the lubricant other than the viscosity and the properties of the surfaces.

Carbon Residue Test. Measures the amount of carbon residue remaining in an oil after the oil has been subjected to extreme heating in the absence of air. This will determine the amount of carbon deposit a petroleum oil will form under coking conditions.

Cavitation. Formation of a void because of reduced pressure in lubrication grease dispensing systems; it can lead to failure of the grease to flow to the suction of the system.

Channeling. A term used in connection with lubricating grease to describe the tendency to form a channel by the working down of lubricating grease. In some instances, this can be desirable (as in a bearing), leaving shoulders of unworked grease that serve as a seal and reservoir.

Cone Penetration. Penetration with respect to a lubricating grease is the depth (in tenths of a millimeter) that a standard cone penetrates a sample of the grease under prescribed conditions of weight, time, and temperature. It is a measure of the consistency of a grease to determine its plasticity.

NLGI Grade	Relative Consistency	ASTM Penetration
000	Very fluid	445-475
00	Fluid	430-460
0	Semifluid	355-385
1	Very soft	310-340
2	Soft	265-296
3	Semifirm	220-250
4	Firm	175-205
5	Very firm	130-160
6	Hard	85-115

Figure 9-8.

Typical Penetration Results.

Demulsification. The ability of an oil to separate from water. This can be measured dynamically or statically. This is particularly important in steam turbines.

Dropping Point. The temperature at which a grease passes from a semisolid to a liquid state. This is a quantitative indication of the heat resistance of grease where a semisolid lubricant is required. A general rule of thumb is to stay 100° below the dropping point for continuous duty, and 50° below the dropping point for intermittent duty.

Emulsification. The mixing of water in oils. Except in special cases, such as soluble oils used in metal cutting processes, emulsification is undesirable. It (1) permits oil to retain water particles that ultimately lead to lubricant failure, (2) displaces oil in frictional areas causing mechanical failure, and (3) causes rust and corrosion; reservoir foaming also results in some cases.

Extreme Pressure (EP) Lubricants. Lubricants that impart to rubbing surfaces the ability to carry appreciably heavier loads than ordinary lubricants without excessive wear or damage. These are measurable with many methods such as Timken OK Load, Fourball Method, Falex Tester, etc.

Fire Point. The fire point is the lowest temperature at which an oil ignites and continues to burn for at least five seconds.

Flash Point. The flash point of an oil is the lowest temperature at which it gives off vapors that will ignite when a flame is passed over the surface of the oil.

Fluid Friction. Friction due to the flowability of the fluid itself.

Gravity. There are two types of gravity measurement: (1) specific gravity is the ratio of the weight in air of a given volume of a material at a stated temperature, to the weight in air of an equal volume of distilled water at the same temperature; and, (2) API gravity is an arbitrary scale chosen by the American Petroleum Institute, in which the specific gravity of pure water is taken as 10. Liquids lighter than water have values greater than 10 and liquids heavier than water have values less than 10.

Hydrodynamic Lubrication. The state of lubrication in which the shape and relative motion of the opposing surfaces causes the formation of a continuous fluid film under sufficient pressure to prevent contact between those surfaces. It is commonly called fluid film lubrication.

Incompatibility. Two lubricants show incompatibility when a mixture of the products shows physical properties or service performances that are markedly inferior to those of either before mixing.

Micron. A millionth of a meter or 0.0000394 inch.

Neutralization Number. A number expressed in milligrams (Mg) of potassium hydroxide required to neutralize the acid in one gram of oil. It is used to show the relative changes in an oil under oxidizing conditions. It measures development of injurious products in oils over a specific period of time.

N.L.G.I. Number. A numerical scale for classifying the consistency range of lubricating greases and based on the ASTM D 217-65T penetration number (see Figure 9-8 under cone penetration).

Non-Newtonian. A fluid that does not fall within Newton's mathematical definition of viscosity, but rather an apparent viscosity whose quantitative value may vary widely with varying shear rate.

pH. This is a term used to express the degree of acidity or alkalinity of solutions. Values of pH run from 0-14, with 7 indicating neutrality, and numbers less than 7 indicating increasing acidity. Numbers more than 7 indicate increasing alkalinity.

Pour Point. The pour point of an oil is the lowest temperature at which the oil will pour or flow under prescribed conditions when it is chilled without disturbance at a fixed rate.

Saponification. A process in which a fat (or other compound of an acid) reacts with an alkali to form a soap. By adding an oil to this soap, greases are made.

Semifluid. Any substance having the attributes of both a liquid and a solid.

Sludge. Insoluble material formed as a result of either deterioration reactions of an oil or contamination of an oil or both.

Soap. A compound formed by the reaction of a fatty acid with an alkali.

Soluble Cutting Oil. A mineral oil containing additives that enables it to be mixed easily with water to form a stable emulsion for use as a cutting oil.

Stick-Slip. A relaxation oscillation usually associated with variation of the coefficient of friction, with relative velocity or with duration of static contact. It is also associated with the formation and destruction of interfacial junctions on a microscopic scale.

Synthetic Lubricant. A lubricant produced by synthesis rather than by extraction or refinement.

Tacky. A descriptive term applied to lubricants that are particularly sticky or adhesive to metal surfaces.

Thixatropy. The property of a lubricating grease that is manifested by a softening in consistency as a result of shearing, followed by a hardening in consistency that starts immediately after the shearing stops.

Viscometer or Viscosimeter. An apparatus for determining the viscosity of a fluid. Many such devices are available from small hand units that compare the speed of the fall of a ball in a fixed fluid compared to the fall of another ball in the fluid being measured, to complicated devices that heat the fluid to be measured to a given temperature and then measure the time for a given volume of fluid to flow through a given aperture.

Worked Penetration. The penetration of a sample of lubricating grease immediately after it has been brought to 77°F and then subjected to 60 strokes in a standard grease worker (ASTM D217).

10

Effective Maintenance of Filtration Systems

C.H. Gordon

SCOPE OF CHAPTER

The filter is the first line of defense in protecting heat transfer and building surfaces. This chapter describes the types of air filters commonly encountered in heating, ventilating, and air conditioning systems; how they are classified; what their purpose is intended to be; the limitations of each; and reasons for selection of each type. Instructions are provided for servicing each type of filter. How to maximize economy and utility in filter systems is discussed in detail for each element of cost, with common-sense rules for maximizing overall system life and economy in filtration systems.

Case histories are presented for upgrading filter systems to show the economics of using modern technology. Charts of standard times for testing pressure resistance across filter banks and for changing various types of filter media are included. Finally, this chapter will explain what to do when filtration has not been maintained effectively, including remedial cleaning of hot deck and cold deck coils, plenum chambers, duct work, and diffusers.

COMMON FILTER TYPES

Filters in common use may be classified in several ways: by efficiency, by method, by use, or by construction. Here are the various classifications commonly found:

1. Efficiency

    a. Low = 20 percent or less dust spot efficiency (The American Society of Heating, Refrigerating, and Air Conditioning Engineers Standard 52-76)

    b. Medium = More than 20 percent, less than 90 percent dust spot efficiency

    c. High = More than 90 percent dust spot efficiency, or more than or equal to 95 percent DOP Test (0.3 Micron Smoke)

2. Method

    a. Inertial Impingement: Particles are trapped on media fibers by being impinged by the force created by their weight and high velocity. An adhesive coating holds the accumulated dust in place.

    b. Interception: Particles too small and too light in weight to be impinged are removed most economically by interception, which occurs when their path is altered by air molecules after their velocity is slowed by passing through media. Filter media used in interception is pleated and usually of finer fiber than inertial impingement media.

    c. Electronic Agglomeration or Electrostatic Precipitation: This filter works by electronically charging dust particles so that they will collect on oppositely charged plates. When sufficient small particles bind together or agglomerate, they break away from the plates and are collected downstream, usually on inertial impingement or impingement-interception type media. Older designs require washing in place of collector cells to remove dust and oiling of plates for reconditioning.

3. Usage

    a. Prefilters: Protect both coils and final filtration systems by removing larger dust particles and contaminants upstream, usually right near outside air dampers. Prefilters are usually a panel-type using inertial impingement method.

    b. General Ventilation and Air Conditioning: Includes low, medium, and high efficiency ranges, dry, viscous-treated, washable, and disposable media types, in a variety of frames, construction, and arrangements. Economy in first cost and life cycle is the usual criterion for selection.

    c. Downstream High Efficiency Particulate Air Filter Beds: Designed to meet requirements of Department of Health, Education, and Welfare Resources Publication No. 76-4000 for Hospital and Medical Facilities. Efficiency of 90 percent minimum is required for most installations, to protect the environment of patient care, treatment, diagnostic and related areas, and sensitive areas such as operating rooms, delivery rooms, recovery rooms, and intensive care units.

    d. Industrial (nonatmospheric dust): Includes special media and filter arrangements to remove lint, press ink mist, and other nonatmospheric contaminants.

Classification of Filters by Construction

Panel Media with Coarse Fibers: Usually called throwaway filters. This construction is the common type found in prefilters. The medium most common in disposable panel filters is glass fiber, although polyester fiber is also used. The fiber is treated with an oil or adhesive spray to create a viscous impingement medium. This is usually found in an efficiency range of 20 percent or less (ASHRAE average with atmospheric dust).

A variation of this type of construction is the renewable pad of glass fiber in a permanent metal and wire frame. When the pad collects its full dust load, it is discarded and replaced with another inexpensive pad.

Pleated Media with Less Coarse Fiber, Few Pleats: There are typically two distinct layers of media material having different fiber sizes and packing densities. The less dense, coarse fibered upstream layer removes large, heavy particles, while the finer fibered, more dense downstream layer may also be treated with a special adhesive on the air-leaving side to prevent blowoff of collected dust. The pleats are generally supported and held in place by wire retainers or welded-wire fabric. Efficiency will range from 25 to 40 percent, according to various manufacturers.

Pleated Media with Fine Fibers and Precisely Spaced Pleats: Efficiencies from 50 to 95 percent are available in this type of construction, depending upon the media fiber size, thickness, and dispersion. Capacities for operation in cfm vary with filter media area (net effective) provided. Although generally found in supported media construction, there are nonsupported pleated media filters in this range of efficiencies. Nonsupported media should never be used in variable volume air systems nor in areas of critical care where media blowout could cause contamination.

HEPA-Type Finely Fibered Media: High efficiency particulate air filters were developed originally for clean rooms and gauge and metrology labs used in military and aerospace hardware production, which explains why the military standard testing procedure (M16-STD-282) is used for this type of filter. HEPA filters are now used in pharmaceutical manufacturing, food and beverage processing, surgical and other hospital applications, electronics and aerospace assembly, photosensitive film production, nuclear uses, and many other areas. The distinguishing construction features are high ratio of filter area to face area, and fine fibers with tightly controlled spacing, closely pleated. Performance characteristics typically include an efficiency of 99.97 percent on 0.3 micron size particles at an initial resistance of 1.0"/w.g. and final resistance of 2.0" to 3.0", depending upon construction.

Roll Media: Although typically furnished as a prefilter or upstream filter for a higher efficiency filter system, the roll media filter may be used where 25 percent average efficiency will suffice. The usual roll media installation is an automatically advanced roll of adhesive-coated glass fiber filter media that is fed into the air stream and rerolled after it has collected its dust load. The roll may be advanced by a signal from a timer (at a predetermined rate), or better by an inclined draft gauge to advance the roll only when the design final pressure drop is reached. Manual crank models are also available.

Side Access Housing: Packaged, side access housings are available for combinations of almost all types of prefilter and final filter construction. The advantages include ease of access for servicing, installation where limited headroom is a factor, and ability to service from outside the duct, usually from either side.

Activated Carbon Filtration: Activated carbon will absorb up to 50 percent of its weight with odors and retain them in the network of tiny pores within the body of the carbon. One pound of activated carbon (approximately 50 cubic inches) contains an estimated six million square feet of surface area. Activated carbon filtration use is limited to odor control in air conditioning systems, although its widest usage is in the chemical processing industry where activated carbon is used to purify gases, to separate gases, to recovery solvents, and to carry catalysts. Activated carbon filters in environmental control systems are usually constructed of trays or panels of activated carbon and placed in the final filtration air stream in a housing designed either for front or side access to the trays or panels.

Electronic Filters: Two types are generally found in HVAC systems: (1) the agglomerator-type with disposable collection media, or (2) the precipitator/collector-type, which requires washing and renewal of collector plates, automatically or manually. The principle of electrostatic precipitation is the same—dust and smoke particles are given positive or negative static charges by the electrostatic field set up by the charged ionizing wires and the grounded struts. Charged particles then enter the collecting section, which is made up of alternately positive charged and grounded plates. The charged particles are attracted to and held by the oppositely charged plates. In the agglomerator-type with disposable collection media, particles build up on the plates until they break off in larger chunks called agglomerates and flow downstream to be captured by the final collecting (disposable) filter. In the precipitator/collector-type, which requires washing and renewal, an automatic or manual wash cycle is required to remove the collected particles to prevent their unloading and passing downstream.

Automatic Vacuum Renewable: This system is not designed for ordinary HVAC installations but for specific industrial situations, primarily laundries and textile mills, where a stable interior environment is essential and large quantities of conditioned air must be recirculated. The system consists of a lint filter that traps airborne particles. An automatic vacuum or pneumatic sweep removes the particles through a manifold to a secondary collection point for recovery or disposal.

UNDERSTANDING EFFICIENCY TERMINOLOGY

Three performance test procedures for general HVAC air cleaning devices are in current use:

- Atmosphere Dust Spot Efficiency

- Synthetic Dust Weight Arrestance
- DOP Smoke Penetration Method

The first two procedures are covered by ASHRAE standard 52-76 (which supersedes 52-68). The third is prescribed by a U.S. Government standard, MIL-STD-282. Much past confusion has been eliminated by standardizing terminology. In the ASHRAE standard, the term *efficiency* now applies only to tests made by the dust spot procedure on atmospheric air and its contaminants. Results of tests measuring *weight* of an injected synthetic dust removed by the air cleaning device are now reported as *arrestance*. The DOP Smoke Penetration Method, which reports results in *count percent,* is designed to distinguish between filters whose air cleaning efficiency exceeds 98 percent.

Of primary importance in building maintenance is the ability to reduce staining by trapping small particles found in atmospheric dust and contaminants. This is the performance measure meant by *efficiency* in the dust spot procedure and is the rating commonly spoken of in classifying filters. High Efficiency Particulate Air Filters (HEPA) are found in industrial "clean rooms," operating theaters, and other areas of hospitals, as final filter beds to meet critical requirements.

The relationships among particle size, efficiency requirements, filter design, materials, and construction are illustrated in Figure 10-1.

SERVICE PROCEDURES FOR AIR FILTRATION SYSTEMS

The general rule for servicing environmental air filters among too many maintenance people has been:

Look at it occasionally.

If it looks too dirty, clean it or change it.

Unfortunately, this approach ignores all principles of engineering economy. Let's look first at some guidelines for maintenance that apply to *all* filter systems, then to specific procedures for each type.

Filtration system economy is based on trade-offs among three major cost components: filter media, labor, and energy. It is easy to see that *filter media cost* would be minimized by using the cheapest media for the longest possible time. The labor cost component, however, has three elements: labor to service filters; labor to clean coils, ductwork, and registers in the distribution system and labor to clean building ceilings, wall surfaces, and furnishings. While the strategy that minimizes filter media cost will also minimize service labor cost, the likely effect will be to increase the cost of labor to clean air distribution system internals and building surfaces. Many filters have a tendency to *unload* excess contaminants once their capacity has been reached. They discharge downstream, fouling heat exchange surfaces, coating the inside of distribution ducts, and spilling into conditioned spaces, showing up as sooty soil on ceilings, diffusers, drapes, and wall surfaces.

Energy cost is affected by two elements related to filtration maintenance—heat

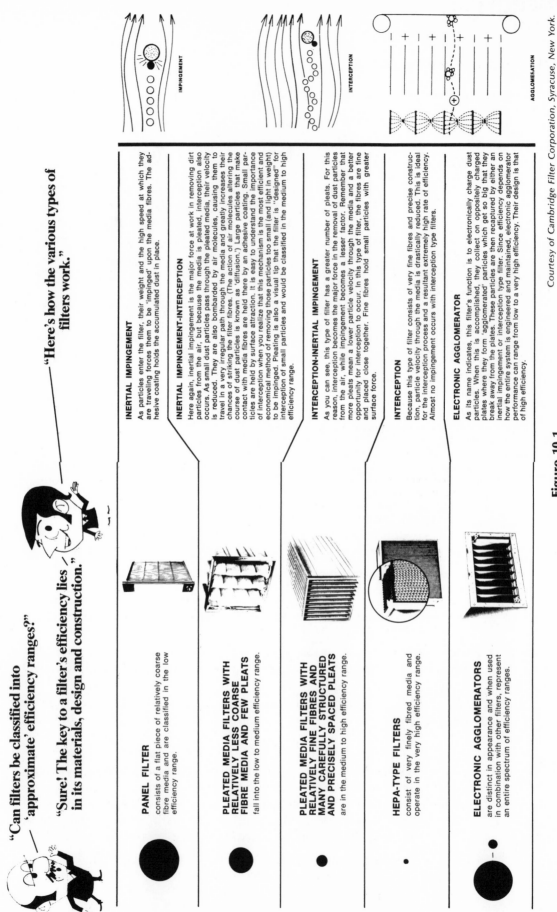

Figure 10.1.

Various Types of Filters and How They Work.

Courtesy of Cambridge Filter Corporation, Syracuse, New York.

328

exchange at coil surfaces and static pressure on the fan caused by resistance to the air stream. Optimum energy costs demand the best possible heat transfer between the coils and the airstream. The work required of the fan is minimized when the least resistance to the airstream is presented. Since the filter's resistance to air flow increases as it does its work of removing contaminants, allowing the filter to "load up" to its design capacity may create extra energy cost[1] in excess of the cost of additional service labor and filter media. A good case can be made for changing filters at approximately 80 percent of design final resistance,[2] since the energy cost rises sharply with resistance and the first 80 percent of resistance (and filter life) takes a much longer time period than would be proportional to resistance.

In order to optimize total system economy, the engineer would need to evaluate the fan characteristics (volume, static pressure, velocity pressure, and horsepower required) and the filter characteristics in order to determine design, initial and final resistances, efficiency, and downstream requirements.[3]

The following footnotes and Figure 10-2 were contributed courtesy of John J. Haarhaus.

---

[1]Added energy cost can be attributed to the filtering system if fan speed must be unduly increased to overcome an uncontemplated increase in air flow resistance. In the absence of such an adjustment, fan energy will decrease as the filter loads up.

[2]Reduced air volume is the controlling criteria for filter media replacement. "Design final resistance" defines the limit of a filter's ability to perform. The ability and/or desirability of subjecting the system to this limit must be verified. (See footnote 3.)

[3]Power is required to operate a media-type air filtering system because:

1. In order to deliver a required air volume the fan must be able to overcome the obstruction that the filter imposes by just being there.
2. The fan must be adjusted for a higher initial air volume to compensate for the reduction (5 percent is typically allowed) that will occur as the filter loads up.

The dynamic relationship of the air filter to fan energy can be dealt with objectively. Two simple measurements, a little basic knowledge, and a look at the manufacturer's data should be all that is required.

A. Measurements: The existing system must be defined in terms of fan speed and total static pressure to provide a starting point for comparison (Figure 10-2 - point D).

B. Basics:
1. A glance at a fan curve will show that within its acceptable operating range, a fan's ability to overcome pressure (at constant speed) increases with a reduction in air volume. (See Figure 10-2 - Line D-A.)
2. The system curves, usually superimposed on the fan curve, represent fundamental fan law, which declares that a reduction in air volume causes a reduction in pressure within the system. (See Figure 10-2 - Line D-C.)
3. The increase in pressure difference of an air filter as it "loads" can be assumed to equal the algebraic sum of the two pressure changes described above (Figure 10-2, A (-) C = BA + BC).

C. Manufacturer's Data:
1. The fan curve, and/or tabular fan data, for the system under consideration is readily available.
2. Any filter manufacturer whose product is worthy of evaluation will provide initial resistance data and state his limits of loading.

With this data at hand, all energy factors attributable to the filtering system can be estimated fairly.

The point at which a filter should be changed can be determined as follows: (Refer to Figure 10-2.)

a. Look up the static pressure at point (A) in the tabular fan data and convert to total pressure (P @ A) by adding the fan's velocity pressure (Vp) at the reduced cfm [Vp = (fan outlet velocity ÷ 4,005)^2].

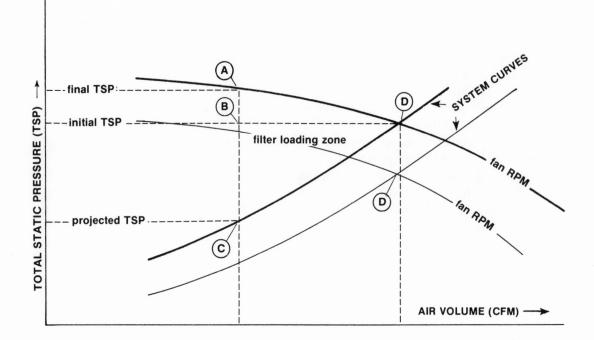

Figure 10-2.

Filter Loading Zone.

(Footnotes continued)

    b. Convert static pressure at point (D) to total pressure (P @ D) by adding the fan's velocity pressure at the initial cfm.

    c. Calculate the system pressure at point (C) by fan law [P @ C = P @ D (cfm @ C ÷ cfm @ D)^2].

    d. The change in static pressure that governs filter replacement equals (P @ A) minus (P @ C); i.e., the pressure drop across the filter which indicates that cfm has been reduced to the point where the filter should be changed.

  Note: If the air system's governing change point is below the load limit of the filter, the allowable pressure drop may be broadened by increasing the initial air volume of the system. Such a change should be evaluated carefully from the standpoint of energy, because fan horsepower will increase as the cube of the increase in air volume. Thus, a 10 percent increase in air volume will result in a 33 percent increase in horsepower.

How to Inspect Filter Installation

There are four important items in any filter installation to check:

- *Support of the filter frames.* If the bank of filter frames is not rigid and well supported, it can collapse as the filters load up.

- *Leaks around frames.* If light is showing anywhere between filter frames and/or between the frames and duct walls, caulk these cracks to prevent unfiltered air from leaking by.

- *Fit of filters in frames.* Incorrectly installed filters will also allow air to bypass. Set them in plumb and square.

- *Condition of the media.* Extended surface (bag) filters should be fully open and extended, not pinched shut. If you see any tears, rips, or holes in any media, replace the filter at once.

Tools Required

Any HVAC mechanic servicing filtration systems needs, at a minimum, these tools:

- A good flashlight, preferably with magnetic handle
- Screwdrivers and nut drivers
- Locking pliers
- A good quality manometer (portable)
- Caulking gun
- Duct tape

How to Change Filters (General)

The following steps should be followed in changing filters in most environmental air filter systems:

1. Remove the old filters and set them to one side in the dust or filter housing.
2. Vacuum-clean or brush the holding frames to remove settled dust.
3. Inspect the holding frame gaskets, and replace any that are damaged.
4. Remove the new filters from their cartons and install according to manufacturer's instructions. It pays to read instructions.
5. Check the filter to make sure the media is not caught or damaged. If a filter was damaged during shipment, contact the delivering carrier and file your claim for loss. If you are using bag filters, check to make sure the pleats are free to open fully.
6. Check all fasteners to make sure they are holding filters in place securely.
7. Remove the dirty filters in the cartons that held the new ones. Vacuum the duct floor, close the duct door, and zero the manometer.

Use of the Manometer

A simple inclined gauge device is inexpensive and easy to use. Manometers may be portable or installed permanently at major filter bank locations. In either case, the purpose is the same—to remove guesswork from servicing filters. All filters have design specifications giving both initial and recommended final resistance to air flow. These specifications are usually stated in inches of water. The manometer is installed to measure pressure drop across the filters in inches of water.

Since oil can evaporate from the gauge, every three months or so the manometer should be *zeroed*. This requires the system to be off if the manometer is permanently installed, or at least the tubing disconnected, taking care not to lose oil from the gauge.

When pressure drop reaches the predetermined point, filters should be changed. Since panel-type prefilters have a relatively short life, measure their pressure drop weekly to prevent having them start to "unload" contaminants. For other types of filters, monthly readings should be adequate to tell the rate at which they are loading. Remember, the last 20 percent of resistance comes more quickly.

SPECIFIC INSTRUCTIONS FOR SERVICING

Panel Filters

There are four main types of panel filters: permanent metal, permanent foam, disposable, and replaceable media. The general service procedures apply for each with the following modifications:

- *Permanent Metal Panel Filters:* Unit filter cells of this type use an impingement fluid that may or may not be coated on the cell at the factory, but must be applied to "charge" the filter before it is placed in operation. The cleaning cycle will depend upon the amount of dust in the air, fan characteristics, and the number of hours in operation. Cells are usually cleaned when the resistance reaches 0.25" to 0.50" w.g., but can be operated at 0.75" pressure drop or higher, depending upon the capacity of the fan. By cleaning one-sixth to one-fourth of the cells during each cleaning period, a constant operating resistance can be more nearly maintained, which will give a more uniform air distribution for the system.

  Remove the dirty cells and replace immediately with clean sprayed cells. Clean the cell by washing with a water spray, soaking in a detergent solution, or immersing in a steam-heated cleaning tank. Water temperature is not critical, but a heated detergent solution only should be used in the solution. *Do not* use caustic (high pH) solutions for cleaning filter cells. After draining off dirty water, the cells should be recharged. It is not necessary for cells to be completely dry prior to recharging. Never try to clean filters by brushing off the dirt. This mats the dirt over the front of the cell, making it harder to wash.

For recharging, a pump-up pressure spray (such as the Hudson sprayer) is ideal. Alternate methods are aerosol spray, or shaking and draining dry. The impingement fluid (VIA Viscosine or equal) must coat both faces, penetrating into the depth of the cell. For a 20″ x 20″ size filter, approximately one ounce fluid per half-inch thickness of filter is required.

- *Permanent Foam Panel Filters:* Open-cell foam or synthetic material (polyurethane or polystyrene) is used for impingement-type media in residential and small commercial equipment. This material generally will not have the capacity of permanent metal and will require more frequent cleaning. The foam is cleaned by washing with a hose or tap water. Impingement fluid is not generally used.

- *Disposable Panel Filters:* Fiberglass or polyester fiber woven media are used in paperboard frames to construct a variety of types and thicknesses of panel filters. Typically the media are coated with an impingement adhesive similar to permanent metal, although with a different viscosity than that used with metal. Servicing is simple—follow the general rules for exchanging clean for dirty filters, and discard the dirty filter units.

- *Replaceable Media:* Permanent frames with disposable pads of filter media are popular replacements for disposable panel filters. The media are treated with the same impingement adhesive as the disposable filters. The procedure for servicing is the same as other panel filters except that the frames snap open for exchanging clean pads in place of dirty pads.

Extended Surface Filters

These filters will operate until they are completely plugged without unloading. Actually, their filtration efficiency increases as more and more dust is collected. However, their resistance also increases, eventually reducing air flow. This makes monitoring of the pressure drop across the filter bank most critical. Also, the consequences of a torn or ruptured bag, releasing contaminants downstream, make it imperative to examine carefully the pleats to assure free opening to full extent of the surface. Supported pleat design filters may be treated essentially as disposable panel filters except for the capacity and pressure drop.

SPECIFIC INSTRUCTIONS FOR TYPES OF FILTERS

Roll Renewable Media

Roll media filters come in a variety of media constructions and resulting efficiency characteristics that allow their use as a prefilter, used alone as the principal filtration device, or to retain agglomerated dust from electrostatic precipitators. The distinguishing feature roll media filters have in common is the use of two spools on either side of the plenum, advancing clean filter media as the used media reaches its final resistance, just as film is advanced in a camera after taking a picture.

The media usually is marked for the last few feet of the roll to show that a clean roll will be needed soon. Some models have a media runout switch actuated through an arm resting on the clean media roll. When the supply is exhausted, this switch opens the circuit, stopping the drive motor and turning on a warning light.

To change media rolls, first the old roll must be advanced to run off the remaining media onto the rewind spool. Then the drive switch must be moved to the *off* position. Remove the used roll and store in carton for removal and final disposal. Take the empty spool from the supply side and put it into the rewind side. The new roll will be covered with a paper wrapper. *Leave this on* until the roll is completely in place, as it keeps the media from unrolling as the roll is on a spool under tension.

After placing the new roll in the metal media, cover; and assuring that all pins, slots, latches, etc., are in proper position, remove the paper wrapper and unwind the media to the rewind spool. The media must be spread evenly across the spool and face the duct or plenum with all air seals and retainers in proper place to assure good fit and prevent air leakage. Engage the rewind spool into the drive socket. Install the keeper.

Engage momentary contact switch and test operation by allowing rewind spool to make two revolutions. The filter should now be ready.

Activated Carbon Filters

The primary task in servicing activated carbon filters is to replace the odor-saturated carbon with new carbon. The trick is to know *when* to replace the carbon, since pressure drop and visual methods are meaningless. Waiting until objectionable odors become noticeable is not recommended. Most activated carbon units are constructed of a number of metal trays filled with activated carbon and stacked like building blocks with alternate ends sealed together by a gasket so that the entire surface of all the trays is presented to the air stream. The number of trays required is a function of the amount of carbon per tray and the velocity of the air flow. Typically, as few as six trays containing 2.5 pounds each would be required for 500 cfm at a velocity of 250 feet per minute. A large bank for 36,000 cfm at 500 feet per minute would require 432 trays of 3.75 pounds of carbon each.

One obvious solution is to have extra trays on hand at all times. This allows exchanging a fresh tray for one in service so that it may be laboratory tested to predict remaining useful life of the carbon. Some manufacturers will perform the test at no charge. If the useful life remaining is sufficient, new trays may be exchanged for old, a few at a time so that not all trays must be renewed at once. Carbon, available in bulk or trays, may be returned to the manufacturer for renewing.

Some carbon filter units are front-opening and some are side-opening. In either case, the trays slide out and are resealed by gasketing.

Electronic Air Cleaners

Agglomeration with Disposable Collector Media: The more modern electronic air cleaners for large installations employ the agglomeration principle. This means that dust particles build upon the collection plates until they break off in larger chunks and are carried downstream to the collection media, which may be bag filters or roll media. The pressure drop across the filter bank gives the cue for removal of soiled media. Standard servicing procedures are followed for the particular media type.

Household Size Electronic Air Cleaners: The typical electronic cleaner for household use fits into an air plenum register or duct. The entire collector unit slides out and may be washed in a conventional dishwasher.

Electrical/Mechanical Servicing: Besides the agglomeration collection media, all electronic air cleaners have two components in common that require periodic maintenance: the power pack, and the ionizer section.

The power pack is a rectifier that supplies high voltage DC current to ionizer wires and plates. It should be checked regularly for proper operation.

The ionizer section should also be checked frequently to confirm proper operation. When operating properly, the wires will be surrounded by a corona visible in the dark as a pale blue glow extending the full length of the wires. The corona is evidence of ionization and the absence of the corona may indicate low voltage or dirty wires. Short circuits will show up as arcs and tripping of the circuit breaker. Broken ionizer wires should be replaced immediately. Occasionally it will be necessary to wipe down or brush dust from ionizer wires, struts, and plates.

TIME STANDARDS FOR SERVICING MAJOR FILTER TYPES

Time allowances or standard times, which can be applied with confidence to maintenance operations, are very valuable in planning and analyzing work requirements. The following time allowances are a combination of observed or experienced times for performing filter service work under average conditions. They should be attainable by any reasonably well-trained mechanic with proper tools and instructions.

Filter Type and Operation	Unit Service Time
Permanent Metal Panel Filters	
Remove and replace	0.75 minutes each
Wash, dry, and recoat	4 minutes each
Permanent Foam Panel Filters	
Remove and replace	0.75 minutes each
Wash and dry	2 minutes each
Disposable Panel Filters	
Remove and replace	0.75 minutes each

Replaceable Media/Metal Frames

    Open frame, remove, replace,   1.5 minutes each
    and close

Extended Surface Media Filters
up to 24" x 24" x 36"

    Remove and replace   1-2 minutes each

Roll Renewable Media
up to 8 ft. width

    Remove and replace   45-120 minutes per roll

Activated Carbon Filter

    Remove and replace trays   1.5 minutes each tray

Electronic Air Cleaners

    Agglomerator type: See Renewable Roll Media
    Removable Collection Unit:

        Remove and replace

        Wash in place collection chambers;   4 minutes each
        Wash, rinse and spray on adhesive   30 minutes each

Access Time Allowances

An allowance must be added to the total filter servicing time to provide time for removing fasteners, removing or opening service or access panels, wiping or vacuum-cleaning the chamber, and reinstalling panels and fasteners after servicing the filter.

Some typical access time allowances are shown below.

Window Air Conditioning Unit

Remove, reinstall screws	2 x 5 x 0.7 min. =	7.0 min.
Remove, reinstall panel	2 x 1 x 1.8 min. =	3.6 min.
Vacuum interior of unit	1 x 5.5 min. =	5.5 min.
		16.1 min.

Package Air Conditioning Unit, 3-9 Tons

Remove, reinstall screws	2 x 12 x 0.7 min. =	16.8 min.
Remove, reinstall panels	2 x 2 x 1.8 min. =	7.2 min.
		24.0 min.

Package Air Conditioning Unit, 10-49 Tons

Remove, reinstall screws	2 x 12 x 0.7 min. =	16.8 min.
Remove, reinstall panels	2 x 2 x 2.6 min. =	10.4 min.
		27.2 min.

Perimeter Baseboard Fan Coil Unit

    Lift, reclose panel   2 x 1 x 1 min. =   2.0 min.

Overhead Plenum Chamber (from ladder)

    Remove, reinstall screws   2 x 6 x 0.7 min. =   8.4 min.
    Remove, reinstall panels   2 x 1 x 8.9 min. =   17.8 min.
    Vacuum chamber   1 x 6.0 min. =   6.0 min.

                                                      32.2 min.

Other items should be calculated to add to filter changing and access time allowances to get a total picture of service time requirements for filter installations. These items include allowances for reading and zeroing manometers, travel time to and from filter banks, material handling time, etc.

MODIFYING AND UPGRADING FILTER SYSTEMS

Many existing filter systems were designed for different purposes than those for which the conditioned spaces are now being used. Codes and requirements have changed. Certainly awareness of the costs associated with energy and maintenance practices affecting consumption and conservation has been heightened. For any number of valid reasons, engineers today are taking a second look at their filtration systems to determine whether a modification is feasible to lower costs, upgrade air quality, or both.

A look at some recent case histories will illustrate the possibilities.

Case 1—An Industrial Plant

The typical air handler plenum had a filter section measuring a nominal 4' high x 8' wide, and was covered with Roll Renewable media. The upgrading consisted of replacing the filter section with holding frames for 24" x 24" x 4" extended area (pleated) filters. Although both the original and replacement media are rated as medium efficiency (20 to 30 percent) by ASHRAE 52-76 standards, the replacement filters are 300 percent as efficient on small particles in the 5 to 10 micron range. Consequently, coil cleaning has been reduced from two times per year to once per year or longer. Labor to change filters has been decreased from 1.5 manhours per change to 10 minutes. Filter media cost is a standoff. List price for eight filters 2' × 2' x 4" comes to a few dollars less than the roll. Frequency of changes is slightly less also.

Total results: Improved efficiency, lower cost of labor in both filter servicing and coil cleaning, and lower resistance to the fan, which could lower energy costs.

Case 2—A Large Bank Building

The main air handling units are provided with an oversize filter bank of nonsupported bag filters, which have an average of 50 percent dust spot efficiency. The design velocity was less than 500 cpm at the filter bank. The engineering staff was expecting to experience two to three changes of filters per year at 1" w.g. final resistance. Instead, the first year of occupancy will see most filters changed only once at 0.80" w.g. The main factors appear to be a conservative design of the filter system and an excellent program of housekeeping in the building, preventing the expected accumulation of dust. The housekeeping program includes daily vacuum cleaning of all the traffic areas of the carpets, which are very high quality, tight woven contruction. Very little of the usual "fuzz" from the top fibers of the carpet has been seen during the initial wear-in period.

The engineering staff is planning to add 2" pleated disposable filters in front of the bag filters when changed. This will extend the life of the bag filters at very little expense in labor and materials since the prefilters will be very easily changed and are expected to cost $4.00 each, versus $17.00 each for the bag filters.

WHAT TO DO WHEN FILTRATION HAS NOT BEEN MAINTAINED

Results of a poor service program are expensive and easily traced from maintenance records. It includes dirty coils, poor heat transfer with resulting high energy bills, freezing of direct expansion coils, fire hazards in distribution system ducts and registers, dirt spills at registers and diffusers, etc.

The correction requirements are all many times as expensive as a good service program. Some of the obvious corrective measures are:

- *Cleaning coils and plenum chambers*

  With a flashlight, determine the extent of residual dirt and fouling on the coils. Dust can be removed by brushing, vacuuming, or blowing with compressed air (and then vacuum-cleaning the settled dust). Oily residue, biological contaminants, and fungus will require chemical cleaning. A number of good chemical cleaning products are available for use with a pressure sprayer, which will remove fouling contamination and leave coils bright without damaging the metal surfaces. Rinsing is not generally required.

  Plenum surfaces should be brushed or vacuum-cleaned to remove all loose dust and debris.

- *Cleaning ductwork, registers, and grilles*

  Heavily contaminated air distribution systems constitute a real challenge. Seldom is enough of the ductwork available for conventional vacuum cleaning to remove accumulated dust. The most successful method is to put throwaway filters over openings and to blow the dust downstream with high volume blowers, collecting it with the filters at diffuser and register openings. Dampers may be closed to section off parts of the system so that blowing and collecting will be limited to only a part of the system at a time.

Heavily soiled registers and grilles should be cleaned with the same solution as the coil cleaner.

- *Cleaning ceilings and walls near conditioned air units*

  Try cleaning first with a portable vacuum cleaner, using a soft bristled brush tool on the end of the wand. If smears result or the area will not come clean, use a neutral detergent solution and a damp cloth. Heavily soiled walls should be washed from the bottom up to avoid streaking. Flat paint on walls may not be possible to clean satisfactorily and may require repainting, but a clean surface is required prior to painting.

CONCLUSION—THE PAYOFF

The benefits of a good air filtration service program are definitely worth the investment of management time to get it established and followed through to successful implementation.

- Enhanced building appearance — This definitely affects marketability of tenant space and suitability for other types of occupancy.

- Energy savings — through reduced resistance to air flow, permitting lower fan speed, and resulting in lower horsepower requirement at design air volume.

- Improved air quality — Occupants of the building always benefit from reduction in the carried over contaminants in the building's conditioned air.

- Labor savings — A good filtration service program will reduce the time to change filters, clean coils and ductwork, and reduce corrective work requirements. Servicing on a rational, planned basis always saves over breakdown or catch-up-type programs.

11

Proper Maintenance of Pumps, Fans, Bearings, and Belts

W. H. Weiss

SCOPE OF CHAPTER

Pumps, fans, bearings, and belts are the major workhorses of industry. It is important, therefore, that they be given high priority under planned maintenance. Two reasons may be given to justify this view: (1) the performance of this equipment affects the efficiency of the machines of which they are components, and (2) their failure to perform properly usually exerts a great influence on the quality and quantity of production, or service of the company utilizing them.

Factors to be considered in planning the maintenance of these critical pieces of equipment will be presented in general, where procedures and instruments will be discussed. Specific recommendations for each category are discussed under separate headings.

PRIME CONSIDERATIONS

The critical nature of these equipment components suggests that they should be inspected frequently to assure economic utilization. Frequent inspection pays off in lower maintenance costs since it minimizes shutdowns for repairs and replacements. A carefully organized planned maintenance program will keep equipment productivity high and downtime to a minimum.

The number and type of diagnostic instruments for predicting mechanical failure have been increasing in the last few years. The maintenance of pumps, fans, bearings, and belts can be handled with a few basic tools. Instruments can make it easier to keep track of equipment performance and efficiency, and to diagnose the cause of failure when it occurs. Among the instruments you should have are thermometers, an ammeter and voltmeter, a vibration meter, and a tachometer.

On-stream and in-operation detection and diagnosis of specific machinery problems are desirable because troubles can be found and an answer arrived at before extensive failure occurs. If you can analyze a problem or recognize an impending failure while the machine is on-line, you can schedule the repairs or replacement work. You can also arrange for manpower, tools, and parts to be available for the shutdown. This approach requires, of course, that you be able to measure some characteristic of the machine that reflects its condition. Bearing temperature is a good example of a measurable characteristic; you usually consider a significant increase in bearing temperature a sign of trouble.

An important characteristic common to all operating machines that can be easily and accurately measured to uncover mechanical problems is vibration. Fans, pumps, and other driven equipment have vibration profiles that are normal and acceptable. When a machine's vibration profile changes, mechanical trouble is usually the reason. Imbalance, misalignment, bad bearings, or loose components will cause changes in vibration profile. Each mechanical defect causes its own unique change from which you can identify a problem positively by its vibration characteristics.

An instrument called a vibration meter or analyzer is needed to measure the amplitude, frequency, and phase characteristics of vibration. Typical equipment today consists of a meter, analyzer, and recorder as shown in Figure 11-1. A predictive maintenance program using vibration testing includes three steps: detecting the vibration, measuring, and analyzing it. Detecting is accomplished by making periodic vibration checks on a machine at selected checkpoints and recording the data to serve as a history of the machine's normal profile. Measuring and analyzing the data reveals trends as well as impending failure. It also pinpoints the trouble area and enables you to decide what repairs, if any, should be scheduled.

Vibration analysis can give you valuable lead time to plan and schedule corrective maintenance. The benefits result in increased production, reduced costs, reduced levels of replacement parts inventory, and better utilization of maintenance labor. The biggest benefit is the avoidance of costly breakdowns that result in complete shutdown of production equipment or interruption of critical mechanical services to a building.

PUMPS

Pump maintenance involves keeping the pumps clean, checking on delivery, stopping leaks, checking the sealing system, and watching that bearings are holding up. All accessories and the system piping should be supported adequately and the controls kept free of dirt and in calibration. The best way to keep up on these matters is to use inspection forms in your preventive maintenance program.

Since you may have several types of pumps to maintain, it is important that you recognize their differences. For example, all positive displacement pumps including

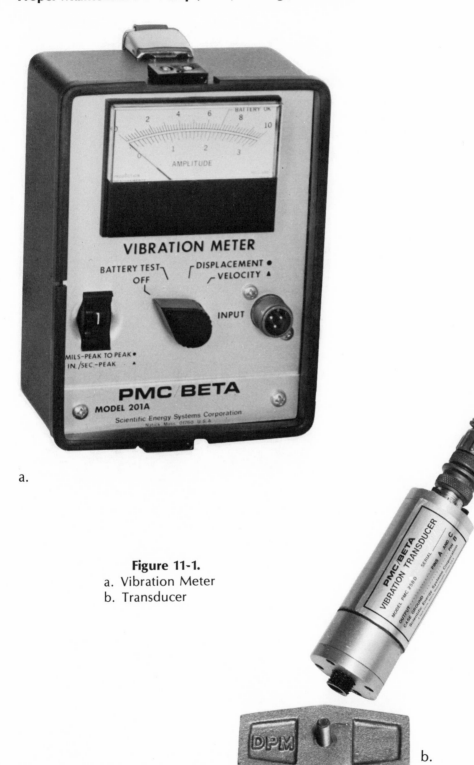

Figure 11-1.
a. Vibration Meter
b. Transducer

c.

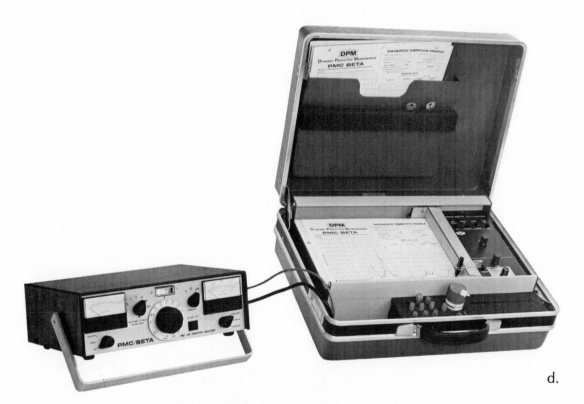

d.

Figure 11-1 (*continued*).
c. Vibration Analyzer
d. X-Y Recorder

Courtesy of PMC/BETA Corporation

gear pumps, piston pumps, vane pumps, diaphragm pumps, screw pumps, and flexible impeller pumps should be equipped with pressure relief valves. Another accessory that must be maintained is a strainer or filter in the suction pipe with the exception of those pumps designed to handle solids in suspension.

Case-in-Point*

Robert Bush, Principal Engineer of The Carborundum Company in Niagara Falls, recently spoke at a seminar on the selection and maintenance of pumps for abrasive applications. He described the company's experiences with centrifugal pumps made from different materials and said that a rubber-lined centrifugal pump with molded rubber impeller was still operating after 16 years on primary slurry service where other pumps had failed within several months. He also recommended the use of air-operated double-diaphragm pumps for pumping concentrated slurries.

The maintenance of pumps is facilitated and minimized if they are installed properly. If you experience excessive failure/repairs with a particular pump that you can't explain, look at the foundation and the alignment with the driver for possible answers to your problem.

The foundation should be strong enough to provide rigid support to the bedplate and absorb strains and shocks in service. A concrete foundation is best. Avoid using unsupported wooden floors for foundation. Faulty alignment causes noisy operation, reduced bearing life, shaft failures, coupling failures, and waste of power. Use feeler gauges and dial indicators to get good coupling alignment.

Packing glands and mechanical seals are the two systems used to accomplish a seal. Both of these systems are affected by friction on the shaft or sleeve, and as such, they require lubrication. Packing should not completely stop leakage—there should be some dripping from the gland—for good shaft packing lubrication. If the packing is tightened excessively, it is likely to burn up. The most common method of supplying lubrication to the packing is by simply bypassing the clean fluid from the discharge volute into the stuffing box. The mechanical seal has several advantages over conventional packing, including limited leakage of product and reduced maintenance.

Figure 11-2 provides a checklist for common pump problems. Figure 11-3 is a checklist for mechanical seal maintenance.

---

*"Chemical Processing Regional Table-Top Show," *Chemical Processing*, June, 1979.

Problem or Cause of Failure	*Solution*
Pump has stopped and will not run.	1. If the drive is an electric motor, look at the overload relays and the circuit breaker. See if the voltage is sufficient. If the drive is an engine, disengage it from the pump and try to start it. If it doesn't start, check the fuel supply, engine oil, coolant, battery, and fuel line. 2. Try to turn the shaft manually. If it turns easily, the failure to run may be due to a temporary loss of power. If it turns with difficulty, there may be an internal part failure or foreign material in the housing.
Too little or too much pressure.	1. Check the pressure gauge to be sure it is correct. 2. Look for an obstruction in the suction piping. Examine the inlet strainer and clean it. 3. See if a valve in the suction line is closed or became partially closed by vibration. 4. Check the speed of the pump with a tachometer or a strobe light. 5. Check the pressure regulator, which may be spring-loaded and adjustable. 6. Examine the pump parts for excessive wear.
Too little or no flow.	1. Look for an obstruction in the discharge piping. Examine the discharge strainer. 2. See if a valve in the discharge or suction line is closed or became partially closed by vibration. 3. Check the speed of the pump with a tachometer or a strobe light. 4. See if the pump has lost its prime. Stop it temporarily to bleed off air. Try feeding liquid into the intake. 5. Examine the pump parts for excessive wear.

Figure 11-2.

Checklist for Common Pump Problems.

Problem or Cause of Failure	Symptom and Solution
Overheating.	Overheating and lack of lubrication are usually related. Stress cracks appear on the hard face surface in direct contact with the carbon mating surface. Other symptoms of overheating are hardened and cracked O-rings. To eliminate such failures, operate the seal in an environment having a lower temperature than the limits of any of its component parts.
Misalignment.	This happens when the seal is installed so that the mating faces are not parallel to each other and perpendicular to the shaft. Be sure the gland is drawn up evenly but do not overtighten it—finger-tight is usually adequate. See that all gaskets and/or O-rings are properly installed.
Abrasive damage.	When abrasive particles get between the seal faces they groove their mirror finish. You must keep abrasive liquids out of the seal housing to prevent this occurrence.
Corrosion.	Corrosion is demonstrated when metal parts are pitted or the sections are totally deficient in metal content. Don't downgrade the metallurgy of the seal from that of the pump. Select the more noble metal for the seal if the application permits.
Lack of lubrication.	Mating faces must run on a liquid film. If there is no film, the seal will overheat and fail. Lack of lubrication at the seal faces commonly occurs when handling hot water or light hydrocarbons. To ensure lubrication and dissipate heat, provide flushing at the seal faces.

*Courtesy of: Pump Care Manual, Goulds Pumps, Inc.

Figure 11-3.

Checklist for Mechanical Seal Maintenance.

FANS

The best way to assure that fans receive proper maintenance is to conduct regular checks and inspections on them while they are operating, and to keep records of your findings. By studying the records, you will learn when to schedule repairs or parts replacement.

Fan inspection requires experience and thoroughness, because wear and material buildup are difficult to detect in the internal areas. Base the frequency of inspection on the amount, type, and corrosive or abrasive nature of the dust entrained in the gas.

The temperature of the bearings on fans should normally remain constant providing that bearing load, speed, and the condition of the lubricant do not change. A sudden increase in bearing temperature or vibration may be a warning of impending failure. Usually the surface temperature on ball or roller bearings should not exceed 190° F; most fans operate at a much lower temperature.

Case-in-Point

An induced draft fan had been inspected six months earlier, but the periodic vibration check indicated that the bearings in the motor were approaching failure. Previous outages had required two weeks downtime and an outside vendor had balanced the rotor. This time the rotor was repaired and balanced in place, and the total downtime was only five days.

Fan bearing vibrations, measured on the top, side, or end of each bearing housing, should be compared with values shown in Figure 11-4 as follows:

Fan Speed (rpm)	Vibration Amplitude (mils)	
	Normal	Don't Exceed
Below 300	0 to 5.0	20
600	0 to 3.5	15
720	0 to 3.0	12
900	0 to 2.5	10
1200	0 to 2.0	9
1800	0 to 1.0	6
2400	0 to 1.0	4.5
3600	0 to 0.7	3

Courtesy of: Wendover, W.E. "Ventilating Fans—Operating and Maintaining," Maintenance Engineering, March 1974.

Figure 11-4.

Fan Bearing Speed Versus Amplitude.

Cause	Diagnosis and Remedy
Worn coupling.	High vibration readings on the inboard bearing frequently in dicate a worn coupling. Replace it and be sure to check it for proper alignment.
Poor alignment.	Misalignment is confirmed by using a dial indicator. Shims are used to reorient the fan and drive for optimum shaft alignment.
Buildup on blades.	Unbalance of fan detected by vibration analysis and confirmed by inspection. A complete internal cleaning and metal repair should be made followed by rebalancing.

Perry, Robert E., "Maintaining Centrifugal Fans," Plant Engineering, June 10, 1976.

Figure 11-5.

Checklist for Causes of Fan Vibration.

Coating	Corrosion and Abrasion Resistance	Max. Tip Speed, fpm	Max Temp. °F	Thickness, mils
Rubber	Excellent for most acids and alkalies. Excellent abrasion resistance.	15,000	180	125–250
Epoxy, epoxy phenolic Teflon S, vinyl chlorates	Excellent to most acids but little resistance to abrasives.	33,000	300	5–20
Plastisol	Excellent to most acids; good abrasion resistance.	20,000	160	80

Perry, Robert E. "Maintaining Centrifugal Fans," Plant Engineering, June 24, 1976.

Figure 11-6.

Characteristics of Typical Fan Protective Coatings.

A checklist for the causes of fan vibration is given in Figure 11-5.

A drop in fan speed and motor amps over a period of time may indicate that belts have stretched or worn. However, be sure to read the air temperature and fan speed when you take amp readings because fan power is proportional to air density and to speed to the third power.

If your fan suffers corrosion from the contaminants in the airstream, you should consider a coating to be applied for protection. Fan housings are usually painted to prevent corrosion, but if abrasive materials are handled, the scroll should be clad with rubber or an abrasive-resistant material. The characteristics of various fan coatings are given in Figure 11-6.

Wear will occur the most on the blades or center plates of fans. If you decide not to attempt repairs, blades should be replaced when wear has reduced them to half of their original thickness.

Wear repair can be handled in two ways. If the fan wheel cannot be removed for repair in the shop, you can apply a metallized coating approximately 30 mils thick. A common coating for this purpose is Carbon 80, which you can buy in wire form and then impact it on the fan blade surface. Wear pads can be applied over damaged areas but they should not be added to a fan without consulting the fan manufacturer. Adding them increases the centrifugal force on all structural members and can cause serious damage to the fan. Only a fan specialist should do this work because it takes specialized skills. Of course, after wear pads are added, the fan wheel must be dynamically balanced before it is put back in the housing.

BEARINGS

The proper maintenance of bearings should begin on the day you receive them, continue through installation, and peak with the lubrication you give them in service. All three of these phases are important to getting maximum life from bearings.

Storage

Many bearings are ruined before they are put in use. Bearings are always packaged and shipped in paper soaked with rust-preventive oil. Don't remove the bearing from this paper until you are ready to mount it. Store the bearing in a dry, clean area, not subject to vibration or stray electrical current.

Mounting

Bearings are often damaged in mounting because of inattention or carelessness. Always read and follow the manufacturer's instructions. Make certain that the housing and shaft are within tolerance. If the bearing is to be heated, use an oil bath or oven—never use a torch directly on the bearing, and never heat one higher than 250ºF. Coat the shaft with a solid lubricant before pressing the bearing in place.

Lubrication

Inadequate lubrication causes many bearing failures. Most bearing manufacturers will tell you of several greases suitable for use with their products,

---

1. Before doing anything, ask if any changes in operation were made. Had the machine been speeded up? Were new belts installed—perhaps too tightly?

2. Check the mountings. Are nuts and other fastenings tight? If the bearing is adapter-mounted, is the adapter sleeve tight on the shaft?

3. If the bearing was pressed on, is the inner race secure?

4. Examine the seals, looking for signs of excessive wear, excessive heat, lack of lubrication, and over-pressurization.

5. Sample the lubricant, comparing it to fresh lubricant of the same type. Is it clean? Has heat or age caused chemical breakdown?

6. Look for signs of poor mounting practice at the edges of the races. Was the bearing "pressed" on with the end of a screwdriver?

7. Shine a bright light on the bearing and wipe the exposed surfaces. Browning, blueing, and blacking indicate progressively higher heat levels.

8. Dismount the bearing. Generalized corrosion may indicate poor storage. Are the shaft and housing also corroded?

9. Inspect the outside of the outer race and the inside of the inner race for fretting. Have the oxides that fretting generates caused rapid wear? Measure the internal radial clearance.

10. Disassemble the bearing. Shiny lines across the raceways may indicate brinelling resulting from shock loads. Was the bearing noisy?

11. Use a magnifying glass on brinelling marks. True brinelling is the imprint of the shock-loaded rollers. In false brinelling, the metal is worn away. Was the bearing subjected to heavy vibration while it was not in use?

12. Stand the rollers on end under a good light. Scuffing indicates that the ends have been breaking the lubricant film. If lubrication was adequate, the bearing was subjected to more thrust than it could take.

13. Follow the wear tracks on the raceways. Crooked tracks indicate misalignment. Are they straight?

Wolfe, Gene, *"Selecting and Maintaining Heavy Duty Roller Bearings,"* Plant Engineering, *April 1, 1976.*

Figure 11-7.

Diagnosing Bearing Problems.

Condition	Recommendations
High temperature	1. Check the lubricant for stability at the high temperature. 2. Change to a journal bearing fabricated in a suitable solid lube material.
Excessive speed	1. Reevaluate the existing bearing. It may be necessary to use a better-class, larger bearing. 2. Check the lubricant for proper viscosity and see that the system has proper scavenging.
High loads	1. Reevaluate the existing bearing. Perhaps use a larger bearing or change to higher strength bearing material. 2. Consider duplex bearings or a larger series bearing.
Very low operating temperature	1. Measure the bearing for proper operating clearance. 2. Check the lubricant for suitability for use at low temperatures. 3. Consider a sleeve bearing constructed with solid lubricant.
Corrosive environment	1. Check the bearing material for compatibility with the corrosive environment. 2. Consider use of Teflon, some other polymer systems, tungsten carbide, and grades of impregnated carbon-graphite.
Abrasive environment or impinging steam or fluids	1. Check for proper sealing to exclude sand and dust from the bearing. 2. Replace a lubricated bearing with a self-lubricating bearing.
Severe shock loading and severe vibration	1. Consider shock mounting the machine to isolate it from the shock and/or vibration. 2. Consider using Molalloy journal bearings which can take high shock loads and operate without external lubrication.

Wolfe, Gene, "Selecting and Maintaining Heavy Duty Roller Bearings," Plant Engineering, April 1, 1976.

Figure 11-8.

Maintaining Bearings Under Severe Conditions.

the quantity required, and the frequency of application. Sufficient grease should be added at each greasing to fill the bearing cavity from one-third to one-half full. Always keep your tools, hands, and work area clean. Oil-type bearings are preferred to grease-type for applications below −40°F or above 350°F. Oil baths should not be

more than half full. Inadequate room for expansion of the lubricant when heated can cause bearing seal failure.

Operation

When a bearing fails, you should investigate why it happened instead of simply replacing it. Figure 11-7 is a checklist of the steps to take in doing this. Maintaining bearings under severe conditions requires that you use a bearing designed for the job. The information in Figure 11-8 was compiled to help you with a problem of this type.

Case-in-Point*

SKF Industries, Inc., bearing products manufacturers, suggests a preventive maintenance program that considers maintenance and operating cost, and optimizes both. Two values are computed for an operation: (1) the cost of an *in-service* bearing failure, C_1, and (2) the cost of a *scheduled* bearing replacement, C_2. Divide C_1 by C_2 to get a ratio. With a ratio of 20 or below, replacing bearings as they fail generally gives the lowest total cost. But at a ratio of 350 or higher, savings can approach 60 percent and more when replacements are scheduled.

BELTS

When V-belt drives are properly designed and properly installed they need very little maintenance. However, drives are sometimes accidentally damaged or the ambient environmental conditions are changed, both of which can cause maintenance problems. To make sure the drives keep working properly, a quick inspection should be part of your preventive maintenance program.

Inspection of V-belt drives consists mainly of looking and listening while the drive is running. Although multiple belt drives run with some variation, all the belts should run at about the same tension. If one or more belts are too loose or too tight, you probably have one or more of the following problems: a worn sheave, improper tension, a damaged belt, and/or improper matching. Figure 11-9 is a checklist for V-belt maintenance. These items discuss the causes and remedies of various V-belt drive problems.

Experts have estimated that 95 percent of all problems with properly designed V-belt drives can be corrected by tightening the belts. You can roughly determine if you have the correct slack in your V-belt drive by measuring the sag of the belts. The correct slack varies with shaft centers and is as follows.

---

*McCool, J.I., "Finding the Optimum Time for Bearing Replacements," Plant Engineering, June 9, 1977, p. 157.

Shaft Center	Sag	Shaft Center	Sag
18 in.	3/8 in.	42 in.	7/8 in.
24 in.	1/2 in.	48 in.	1 in.
30 in.	5/8 in.	54 in.	1-1/8 in.
36 in.	3/4 in.	60 in.	1-1/4 in.

There are three ways to decide whether a drive is properly tensioned under dynamic conditions. First, adjusting by sight entails setting the center distance until the belts have only a slight bow on the slack side when they are operating at full load. Second, adjusting by sound is appropriate for machines, such as fans, that require peak torque on starting. If the belts squeal as the motor comes on, they are not tight enough. Belts will last longer if the squeal is eliminated. Third, adjusting by touch involves measuring belt slippage by the amount of friction heat developed from it. Excessive slippage generates enough heat that with the drive stopped, you cannot hold a finger in one of the grooves. You must tighten the belts until the sheaves run cool.

Case-in-Point*

Jerry W. Wolfe, Application Engineer at the Belt Technical Center, Dayco Corp. in Dayton, Ohio reports, "Proper tension can be easily adjusted. The general practice is to strike the belts with the hand. Under correct tension, they will feel *alive* and *springy*. Loose belts will feel *dead*. Tight belts will yield no give at all."

---

*Wolfe, Jerry W., "Planned PM Pays Off on V-Belt Drives," Maintenance Engineering, September 1975.

Trouble or Problem	Cause	Remedy
Relatively rapid failure; no visible reason.	Tensile members damaged through improper installation.	Replace with all new matched set, properly installed.
	Worn sheave grooves (check with groove gauge).	Replace sheaves.
	Underdesigned drive.	Redesign.
Sidewalls soft and sticky. Low adhesion between cover plies.	Oil or grease on belts or sheaves.	Remove source of oil or grease. Clean belts and grooves with cloth moistened with nonflammable, nontoxic degreasing agent or detergent.
Sidewalls dry and hard. Low adhesion between cover plies. Bottom of belt cracked.	High temperatures.	Remove source of heat. Ventilate drive better.
Deterioration of rubber compounds used in belt.	Belt dressing.	Never use dressing on V-belts. Clean with cloth moistened with degreasing agent or detergent. Tension drive properly to prevent slip.
Extreme cover wear.	Belts rub against belt guard or other obstruction.	Remove obstruction or align drive to give needed clearance.
Spin burns on belt.	Belts slip under starting or stalling load.	Tighten drive until slipping stops.
Bottom of belt cracked.	Too small sheaves.	Redesign for larger sheaves.

"Safety and Practical Maintenance Tips for Your V-Belt Drives," The Gates Rubber Company, December 1975.

Figure 11-9.

Checklist for V-Belt Maintenance.

Trouble or Problem	Cause	Remedy
Broken belts.	Object falling into or hitting drive.	Replace with new matched set of belts. Provide shield for drive.
Noise.	Belt slip.	Retension drive until it stops slipping.
Turnover.	Foreign material in grooves.	Remove material Shield drive.
	Misaligned sheaves.	Realign the drive.
	Worn sheave grooves (check with groove gauge).	Replace sheave.
	Tensile member broken through improper installation.	Replace with new matched set properly installed.
	Incorrectly placed flat idler pulley.	Carefully align flat idler on slack side of drive as close as possible to driver sheave.
Belts stretch unequally, but beyond takeup.	Misaligned drive, unequal work done by belts.	Realign and retension drive.
	Belt tensile member broken from improper installation	Replace all belts with new matched set, properly installed.
All belts stretch about equally, but beyond takeup.	Insufficient take-up allowance.	Check takeup and follow allowance in design manuals.
	Greatly over-loaded or under-designed drive.	Redesign.

Figure 11-9 (continued).

Appendix – 1

Equipment Security in Today's Society

by James H. Thompson, P.E.

The security or protection of plant equipment may take many forms and utilize many types of devices, ranging from ordinary door locks to sophisticated electronic alarm systems capable of producing permanent records of the unauthorized events. From the equipment maintenance engineer's point of view, the protection of his plant, machines, and apparatus must be one of the most important priorities.

For if attacked, however high the odds or low the probability, the damage regardless of degree will be a *fait accompli*. In short, as with other preventive management programs, the success of the security program lies in the advance planning and execution of the preventive measures.

The analytical techniques normally utilized by the equipment maintenance engineer to forecast the useful life of equipment is worthless when forecasting the life expectancy of facilities, machinery, or equipment relative to the probability of incapacitation from willful attack.

What then, must you do to evaluate and plan equipment security maintenance for such indeterminable conditions? To begin, you must realize that a destructive event is possible. Even though it is much easier in today's world to *realize* that a destructive attempt can happen, the plant engineer will still be beset throughout the program by thoughts and comments such as "We've been here years and years and never had any trouble," or, "We aren't a target for criminal break-in because we have nothing of value to steal," or, "Why would anyone want to damage our equipment; it will only lead to layoffs and loss of jobs."

Regardless of these distractive thoughts, security must be effectively and diligently considered. A well thought out and practical application plan must be made for the most likely, most probable, and most unexpected occurrence. You owe it to the owners, management, fellow employees and yourself to apply your best knowledge, experience, and ingenuity to protect the facility for which you are responsible.

Through use of the following approach to an applied security program, you can develop a plan for both personnel and equipment that will provide basic protection and be capable of expansion to fulfill your security needs.

1. Beginning the Program

In very large plants, buildings, or facilities, general protection from the curious, the inquisitive, or the generally nosey public, in the form of fences, walls, and protective grating must be installed. Whether large or small, a plant or building should have designated areas of entry whereby nonemployees are received for business. Provisions may have been made for after-regular-hours monitoring of the premises by a guard or security patrol. In some cases, telephone numbers of management personnel will be posted outside for use in emergencies. A burglar alarm may be installed for unoccupied hours that will report to the police, a private security company, or to a company security department office.

Regardless of the overall security, it is unlikely that the protection of machinery or equipment will have been specifically studied from the view of the plant engineer. All too often top management establishes a plant protection program without realizing the vulnerability of expensive equipment and its potential domino effect upon the business, should it be damaged by an unexpected intentional act. The responsibility from the view of the plant engineer is to begin the protection program by studying and planning methods of protecting buildings, machines, tools, and equipment.

Whatever the size of the facility, the first step in plant equipment protection is to make a close, systematic study of the areas, rooms, lockers, closets, and any location where equipment, tools, or supplies are located. Objectively study each area's contents; consider the potential loss by theft, vandalism, or unauthorized operation. Consider the harassment potential as well as personal injury or damage from unauthorized access. Review insurance and regulatory agency requirements, and safety code requirements for these locations.

Make a list of potential threats. Use your imagination—place yourself as a vandal, saboteur, disgruntled employee, or political activist. Consider the easiest way to disrupt, inconvenience, damage, or steal; consider the resulting impact on occupants, employees, and/or management. Consider the likely damage that may be done by participants of a strike. These considerations will all be of value when you make your final, considered, weighted analysis. You cannot prescribe the cure until you know all the ailments.

Start the list by itemizing the locations and their general contents. Do not try to think of or consider all eventualities at the first attempt, but do make a complete survey of locations. List the security measures and procedures already employed and discuss it with staff members.

2. Physical Security

Doors and Accesses. The most common point of unauthorized entry is through doors. A careful review of doors and accesses to the areas to be secured is required. Inspect each door; the way it was made; whether solid or hollow core; material and condition; and probable resistance to attack. Note particularly the hinge installation, if the hinge pins are accessible from the outside.

Locks. Carefully examine the locks. In many cases ordinary interior door locks may be aborted easily with a screwdriver, knife, or credit card. There are many excellent high-security keylocks on the market. Some adapt to master and submaster systems; others do not. For most multiple area facilities, the convenience of the mastering system is highly desirable; however, remember that key control is essential. Master keys should be held on premises if possible with current responsibility lists maintained.

Keys. Consider hours of open entry, accessibility of keys to the unauthorized, opportunity for key duplication, and keys in the hands of former employees. Keys have served as security for so long that we almost take for granted that the key turning in the lock will assure control of unauthorized entry. A high percentage of theft occurs without the need for criminal break-in. A rekeying schedule could be a valuable deterrent.

Consider the type of bolt lock action currently installed on your access doors. If the bolt may be pushed easily back in while extended, you may assume that the credit card, knife, or screwdriver may also be used easily instead of the key. Consider carefully the results that an intruder may attain by entry and select the lock set accordingly.

Windows, Hatches and Skylights. A window is a very common means of criminal entry because it is often remote, and it is relatively easy to break a small pane and unlatch to gain entry. Entry through roof hatches and skylights may be a direct path into a secured area. Consider the installation of bars, expanded metal screens, or other barriers to prevent such access. A quick, well-planned hit by a vandal can result in major damage before physical response can be implemented, even if it is heard or an alarm is sounded.

Walls. The determined criminal who considers the result worth the effort may consider forced penetration of a wall. In your review of protected areas, do not overlook the possibility of wall compromise, particularly in areas of high value.

Utilities. The protection of service entries, transformer banks, meters, and the like, which provide service to the building, must be reviewed where exposed or outside. Consideration should be given to the installation of a chain link fence, topped by angled barbed wire; for critical areas, the incorporation of a fence or yard alarm may be considered. A reliable fence installation company should be consulted for details of fence selection.

The element of the security analysis is now complete. The potential threats have been analyzed—the areas of attack and the physical constraints, or lack thereof, will provide physical protection. At this point, the plant engineer must have some knowledge of what is available to provide additional protection and thus compromise the threat. General knowledge is necessary, however a security specialist can be retained to do the detailed physical security design.

There are two general classes of electronic sensors. The first is the perimeter and/or point sensor or device, which in effect, makes a line around an area. The second type is the space or area sensor that provides a volumetric coverage. Some sensors by their nature overlap into both categories. For example, a photo beam on an outside wall can be considered a perimeter sensor, while the same sensor deployed across an area can be considered a space sensor. Table 1 lists typical perimeter/point sensors. Table 2 lists typical area sensors.

Table 1: Common Perimeter/Point Sensors	
Magnetic switches	Vibration sensors
Plunger switches	Photo (IR) beams, active
Hinge switches	
Window foil	

Table 2: Common Area Sensors	
Ultrasonics	Audio detection (impulse type)
Microwave	Closed circuit television
Passive infrared	Strain detectors
Audio detection (listen type)	

Alarm systems should be selected by an experienced security specialist or professional alarm company representative. Reliability, freedom from false alarms, required protection and possible expansion can all be possible with proper planning. And the opposite—overprotection and overcost, underprotection and many false alarms can result from improper planning.

Systems can be grouped into electromechanical, photoelectrical, capitance, soundwave reflection, sound listening, microwave, and closed-circuit television. There are advantages and limitations to each system.

The most widely used system is the electromechanical switch. This system functions through electric circuits balanced so that a break in the circuit that stops current flow or a ground that increases current will set off an alarm. Such things as foil strips on windows; magnetic or contact switches on doors; mercury switches on tilting openings; vibration detectors, screens and traps in walls; duct openings; floors; and ceilings comprise this system.

In order to understand the operation of an electrical alarm system, the basic circuit is shown in Figure A1-1.

The basic alarm system shown in Figure A1-1 is typical of that used in many control systems. Relay K1 is the "house" relay. Battery #1 is coupled through R1 and the loop to the coil of K1. The contacts of K1 are connected through battery #2 to an alarm bell. Notice that K1 is energized in the current configuration, and thus the bell is not energized. Assume S1, which is a single-pole single-throw switch, is caused to open. This action stops current from flowing through K1 (the house relay), which in turn causes the relay contacts to change state, which in turn causes

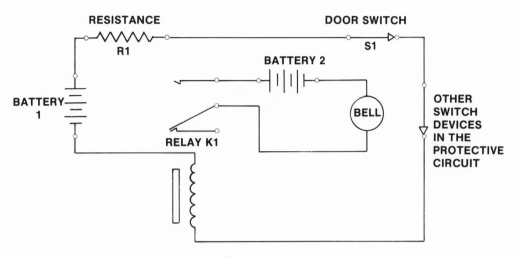

Figure A1-1.

Basic Alarm Circuit.

the bell to ring. This very simple alarm system indicates only the basic principle of an alarm system. This system has many shortcomings and indicates the principle only for closed-loop systems.

Today, almost all effective systems use a basic form of this circuit. Remember that unlike a door bell alarm, the security alarm circuit sounds upon a *break* in the sensor circuit and is thereby self-supervising. If the sensor circuit is defective, the alarm will not set up for protection; or, if it becomes defective while active, an alarm will be initiated.

Sensors other than mechanically activated switches are used in the sensor circuit of a security alarm. However, remember that whatever the sensing technology used, upon receiving an alarm, the device opens a switch that is part of the primary sensor circuit of the basic alarm control unit.

Photoelectric. A light source, mirrors for bending the light beam, receivers, and control devices make up this system. Older systems utilized infrared filters to make the beams invisible. More modern technology uses light-emitting diodes that radiate infrared. Interruption of the beam will set off the alarm. Infrared beams are often used to lace banks of windows or large overhead doors.

Sound Wave (Ultrasonic). These systems utilize transmitting and receiving units in the ultrasonic range, approximately 25 to 46 thousand hertz, and the Doppler effect to produce an alarm indication. Any movement of a body within the protected area causes the receiver to detect a change in frequency and is interpreted in the receiver as an alarm. Permanent as well as portable units are available for installation.

When installed and properly balanced, this is one of the most effective systems in use today. Balancing must be performed to provide the best coverage in that all types of furnishing will affect ultrasonic coverage. Such systems should always be balanced after the room or area to be protected has been furnished. Ultrasound will not generally penetrate doors, windows, or even curtains.

Common applications are vaults, laboratories, tool rooms, record-retention centers, and warehouses. Caution must be exercised when applying ultrasonics to protect a room with moving or rotating machinery, air compressors, steam valves, and other sources of modulated or broad band ultrasound. Such sources may trigger false alarms.

Microwave (Radar) Units. A variation of the Doppler ultrasonic consists of a unit fitted with an antenna that saturates a room with an electromagnetic field. Changes in movement within the area will cause an alarm. Care must be exercised in using this system and in its adjustment, since it will penetrate walls, which can be both advantageous and disadvantageous.

Capitance (Proximity Alarm). After being properly balanced, the capitance system sets up an electrical field around the object being protected. Introducing an additional body into the field will change the capitance and initiate an alarm. These systems are often used to protect safes and groups of filing cabinets. It is important that the safe or cabinets be insulated from ground for this system to work.

Sound Sensing Devices. There are two general classes of sound sensing devices. Both consist of microphones and amplifiers. At this point they diverge. The listen-in sensors connect speakers at a remote site to the amplifiers, and, in effect, someone has an extension of his ears in the area to be protected. The second system or impulse sound-sensor connects a filter and counter to the amplifier. An alarm is generated when a preset number of impulses within a time frame occur. The second system is generally used in vaults that one might attack with hammers, jack hammers, drills, etc.

Fire, Smoke, and Water Sprinkler Alarm Systems. Although primarily installed to provide alarming and extinguishment of accidental fires, the systems work equally well in the case of arson. For areas that already have sprinkler systems, a water flow switch interconnected to the fire alarm system will provide immediate alerting should a sprinkler head open. This water flow switch connection is equally effective preventing damage to building contents, for it is a recognized fact that water damage from a sprinkler system often exceeds the fire damage.

For buildings without sprinklers, it may be desirable to install either a full or partial water-sprinkler system. Although a considerable outlay is required initially, in many cases the long-term savings in insurance premiums will cause a favorable decision.

As an alternative to a sprinkler system, a properly designed heat and smoke detection system may be considered for early-warning fire and smoke detection. Great improvement and progress in the early fire and smoke detection sensors have been made in the past few years. Heat detectors depend on the expansion of a bimetallic strip to close a switch. Smoke detectors are generally of two types: the ionization chamber unit, and the photoelectric sensing unit. Both types have weak points of performance; however, selection and application of the proper type by an experienced fire alarm system designer should provide effective, early warning of an unsafe fire condition.

Standby Power. All systems should be connected to emergency power sources so that operation will continue even in the event of power failure. A simple battery system will usually suffice; more elaborate systems may require motor-generated or other AC systems.

Monitoring Stations. When installed, the security system may be terminated at a guard or night watchman's station at the protected facility. If no guard is on site, the system can be connected by leased telephone lines or a digital communicator to a police department or private monitoring station. The use of direct wire lines provides more reliability and is supervised upon a loss of signal from the protected premises; however, the monthly expense will be more because of the line charges.

A digital communicator connected to one of the facility's regular telephone lines will send an alarm signal to the monitoring station through the use of complex dialing and receiving techniques. Since the line is used only when an alarm is being sent, no monthly line lease charges are incurred. A small one-time charge is made by the phone company for making the interconnection to the alarm communicator.

If you require monitoring service, consult a reputable company in your area to learn of the best-suited options to providing alarm installation and signal monitoring.

Closed Circuit TV. Among the devices being used for security control is closed-circuit TV (CCTV), consisting of a camera placed at a strategic location to observe either inside or outside a protected area. In addition, CCTV has been used to monitor smoke stack emission, cooling tower operation, and console gauge readings. When applicable, the system performs well; however, the cost must be justifiable and the monitor in most cases must be viewed by a guard or assigned person. The biggest advantage of CCTV surveillance comes from the use of a videotape recorder (VTR) to capture for replay and identification the unauthorized events. A video system is a sizable deterrent to would-be intruders, since they are never quite sure when they are being watched.

Several new products have been introduced recently as adjuncts to CCTV that will make multiple variations in application possible.

1. A video storage circuit that will activate a monitor, ring an alarm, or start a VTR upon a change of fixed field of view. This literally means that a switch can be activated when, for example, a person enters the area of a room that a camera is focused upon, thereby making continuous personal monitoring unnecessary.

2. A signal processor that can transmit a slow-scan video picture over an ordinary telephone line. This can be used for unattended locations at a remote distance from the monitoring station.

3. Several types of badge, signature, or identification magnified cameras for special high security applications.

The above description is to act as an idea source to stimulate the imagination for solutions to problems. In no case is a do-it-yourself approach to CCTV application recommended. An experienced, competent application company should be contracted to apply your ideas to hardware.

Electronic Combination and Card Reader Access. A number of manufacturers offer a keyboard push-button panel for access to security areas. By punching in the proper combination, the door is released. In addition to the push-button access control, embossed or magnetically-encoded card readers are available for door control. Some manufacturers offer a combination application of both push-buttons and card readers. The degree of complexity of the system should be tailored to vary widely along with expense and complexity of the system. Accumulate your requirements and ideas and get a good applications company to help.

3. Summary

For your review, the following outline is offered as a checklist for your security program.

a. Be aware of the possibility of either wanton or willful attack on equipment and facility.

b. Study the areas and equipment to assess the damage or loss potentials— use imagination.

c. List the security measures and procedures already employed.

d. Plan the additional security measures required to provide the level of security desired.

e. Determine whether a monitoring station on premises is required or whether remote monitoring would be more cost effective.

f. Select an experienced, competent security application company, if necessary, to finalize hardware selections.

g. Establish a budget for the security operations that is consistent with the risk potential.

h. Implement the program, completely or in logical steps, if the total program cost warrants.

i. Finally, check the installed security measures regularly to assure continued satisfactory operation.

4. Conclusion

For the plant engineer, the protection of his buildings, tools, machinery, and equipment is a must in today's society. Under optimum conditions, an unprotected tool room will be "borrowed" (stolen) from. Open accessibility of a machinery room, hose cabinet, supply storage closet, etc, will be an overwhelming opportunity to the thief or disgruntled employee. Willful destruction, theft, or vandalism is no longer a remote occurrence. It will likely become a probability on the occasion of a vengeful strike, a dissident, a dissatisfied work force, or the thief, from the evident pattern of management laxity to fundamentals of security awareness and protection. In the area of equipment, an ounce of preplanning protection can be worth many pounds of regrettable repair.

Appendix – 2

Cost Reduction Through Proper Energy Conservation Management

by Charles E. Marino, P.E.

INTRODUCTION

The management of energy conservation has taken on many forms and has been tackled in many ways over the past few years; and yet we are still plagued with future fuel shortages and higher utility bills. It is important to remember that energy management is everybody's business—from the custodian to the president of your facility.

Because they are directly associated with the plant maintenance function, the plant engineering staff becomes the catalyst toward detecting and solving areas of energy waste. They are also responsible for educating those people who work in the building about energy management.

All you have to do to start saving energy is to look around and take heed of conservation measures. Energy conservation management requires a systematic procedure of building analysis—both inside and out. The key elements to consider are the following:

1. Building structure
2. Electrical usage
3. Cooling systems
4. Heating systems
5. Heat recovery techniques
6. Kitchen operation
7. People

I. BUILDING STRUCTURE

Though made of basic components such as steel, glass, wood, and aluminum, every structure is unique. Look at the positioning of the building. Which elevation faces North? South? Winter weather will certainly eat up energy dollars unless you observe some of the following precautions:

A. Reduce air infiltration as much as possible. This includes roof penetrations, door openings, cracks in walls, and window openings. Seal up all unnecessary openings, add self-closing devices to exterior doors, caulk wall cracks and the area around windows and doors. Repair window frames and replace broken glass.

B. Observe the amount and type of exterior glass. Consider two energy-related items pertaining to glass.

1. "U" factor is the coefficient of transmission expressed in $\dfrac{\text{BTUs}}{\text{hr./sq.ft./°F.}}$

   The higher this number, the greater heat transfer through the glass.

2. "S" shading coefficient is a numerical measure of solar heat gain on a windowed area. The lower the number, the better the material limits solar heat penetration.

   Take for example, ¼" clear single-plate glass versus ¼" bronze double-plate glass

	"U"	"S"
Single-plate glass	1.10	.93
Double-plate glass	0.60	.56

   From this one comparison, it can be seen that the double-plate glass transmits half as much heat as the single-plate glass. In addition, the double-plate glass will limit about half the solar heat penetration as single-plate glass. So wherever practical and possible, convert to insulating double-plate glass.

   Also, consider the following in office areas: If shading-type film is applied to a single-plate glass window and white opaque shades, or a light-colored, closed-weave drapery is added to the inside of the glass exterior, a savings can be realized.

C. Building insulation has been eliminated or reduced from an energy-efficient thickness in the past by architects and engineers alike as a last-minute dollar-saving "superficial" cost.

In many facilities, no practical or inexpensive method is available to decrease the transmission coefficient "U" value of the building exterior. During renovation or refurbishing operations is the time to apply any initial or additional insulation on exterior walls. If this is not taking place, then you must take the following steps:

1. Where available, check as-built architectural drawings to determine the materials of construction.

2. Check an ASHRAE guide to determine the heat transfer coefficients for the various building materials.

3. Determine how much additional insulation is needed on walls and roofs. Don't forget crawl spaces under uninsulated floors.

4. Call in insulation experts to provide application and cost information.

5. Get the job done as soon as possible. Check your eligibility for tax relief via federal energy credits or financial assistance.

II. ELECTRICAL USAGE

A most significant area for energy reduction is the proper use of lighting and electrical equipment. Some items are:

A. Utilize selective lighting throughout. Consider that lighting is used basically for the following purposes:

1. To properly perform visual tasks.

2. For safety purposes.

3. For decorative purposes.

We have tended to overdesign lighting levels. Therefore, check light levels (in foot-candles) throughout the facility and make certain they are not excessive. Task lighting should be used where high levels of illumination are required. The Illuminating Engineering Society (IES) recommends the following lighting levels:

	AREA	FOOT-CANDLES
1.	Perimeter of building	5
2.	Office areas	70
3.	Corridors, elevators and stairways	20
4.	Toilets and washrooms	30
5.	Entrance lobbies	10
6.	Dining areas	20
7.	General manufacturing	50
8.	Mechanical rooms	20

Reduction in lighting levels via the use of lower wattage lamps or elimination of groups of lamps should be made to the minimum acceptable IES levels. Do not, however, compromise safety for dollar savings.

All decorative lighting both indoors and out should be re-evaluated and all exterior lighting reduced to minimum levels necessary for safety.

B. In addition to proper lighting levels, lamp efficiency should be considered. Lamp efficiency is measured in lumens per watt. A higher lumens per watt performance rating indicates a more efficient lamp. The average lumens per watt for commonly used lamps are:

| | Lumens Per Watt | |
	Including Ballast Losses	Lamp Only
Incandescent	N/A	11-18
Tungsten-Halogen	N/A	20-22
Fluorescent	58-69	70-73
Mercury-Clear	37-54	44-58
Mercury-(Phosphor coated)	41-59	49-63
Metal Halide	65-110	80-125
High Pressure Sodium	60-130	83-140
Low Pressure Sodium	78-150	131-183

Mercury vapor lamps use electricity about three times as efficiently as incandescent bulbs.

Fluorescent lamps are about 4¼ times more efficient than incandescents. Thus it is suggested to convert to a more efficient lamp where possible. Be sure to include the cost of new or additonal wiring, ceiling renovation, and fixture changes when anticipating a change.

Where only incandescent bulbs can be used, install energy-saving types; for example, a 93W bulb in lieu of a 100W bulb.

To ensure continued long-term savings, also include the following lighting-related items:

1. Programmed light fixture and lamp cleaning.

2. Group relamping.

3. Installation of photocells and/or timers.

4. Review ballast operation.

5. Utilize natural lighting wherever possible.

6. Turn off lights when space is not occupied. Self-sticking "reminder" covers that fit over switch plates telling the user to *Turn off lights - Save Energy,* can be obtained free of charge from your local utility company.

C. Stagger electrical loads and arrange work schedules to lower the electrical peak load. Remember that electrical costs are based on consumption and demand. Demand is the peak usage during a specified billing period. This peak is set when a surge of electrical loads are applied at the same time or within a short interval time frame. These loads may be from refrigeration equipment, fans, pumps, boilers, chillers, lighting, and kitchen equipment. When all are started at once, the equipment will peak the electric demand charge.

When electric power demand exceeds the previous peak even for a few minutes, a new demand charge is established on the electric bill. Even though subsequent demand may not come close to that peak again, the new demand charge may prevail for many months to come. Check your local utility company for the applicable rate schedule.

Therefore, it is only common sense to schedule load start-ups so that the demand meter is not subject to unusual demand. *Shedding* of the load can drastically reduce the electric demand and in turn, save money on the electric bill.

Load shedding can be accomplished by electrical timers on specific pieces of equipment or by more elaborate demand controllers. The demand controller basically relates the actual energy rate to a predetermined maximum rate of use. If, during the demand interval, the actual rate approaches a point of exceeding the maximum set rate, selected equipment or loads will shut off (shed) to limit costly peaks.

Another approach used to reduce demand charges is to utilize existing standby generators sets to *peak shave.* By operating the emergency generators, electrical demand will be reduced during peak periods.

III. COOLING SYSTEMS

One of the greatest load consumers is air conditioning and refrigeration equipment. The approach to curtailing this energy usage area is a simple one. Increase building temperature settings to the highest practical and comfortable level.

A summer space temperature of 78°F is judged to be reasonably acceptable to the occupants as well as energy efficient. After building shutdown, the temperature can be further increased to 80-82°F. Take care that this temperature does not adversely affect any equipment or materials left in these areas. When cooling, each degree Fahrenheit that the thermostat is raised will result in about a 5 percent reduction in electrical consumption.

Also investigate the effect of reducing operating time of cooling equipment during the course of the year. Many facilities, for example, specify a cooling season from May 15 through September 15. This period should hold firm even though a few days of discomfort may result. Hours of operation may also be a factor during the day.

Cooling with outside air, whenever possible, should not be overlooked. Once the outside air temperature goes below 65°F, comfort cooling should come strictly from outside air ventilation. Limiting exhaust and air makeup requirements to their minimum factors will also help save energy.

Emphasis should be placed on checking air filters at least monthly and cleaning or changing them when needed. A simple differential pressure gauge installed across the filter will give a quick indication of the static buildup at this point.

Proper water treatment of cooling water is required to maintain scale-free heat transfer surfaces. Scale buildup of only one-thousandth of an inch in an air conditioning condensor equates to a compressor working 11 percent harder.

Many air conditioning systems are being retrofitted with day/night thermostats. These thermostats are controlled by a built-in time clock which, at a predetermined hour, starts and stops selected fans and opens or closes automatic valves to maintain desired temperatures.

Finally, in the area of cooling, you may want to look into *enthalpy* controls to add to air intake fan systems. These up-to-the-minute controllers allow each fan to take maximum advantage of outside air for free cooling.

Solid-state logic circuiting is used to compare temperatures and moisture content of both outside air and building air at all times. Units then decide which air source to utilize for minimum mechanical cooling. It is generally justified to use enthalpy controls in air handling systems in excess of 10,000 cfm.

IV. HEATING SYSTEMS

The most obvious approach to energy savings in the realm of heating is the reduction of building temperature to 65°F. For each degree the thermostat is lowered during an eight-hour period, a 1 percent fuel savings can be achieved.

As with the cooling cycle, automatic resetting of the temperature at night can be accomplished via the installation of clock thermostats.

Similar to electrical demand, if your particular facility uses city steam, check bills for a steam demand factor. If you are being charged for same, closely monitor this factor and adjust steam usage to increase the load gradually, and in turn minimize the demand charges.

Insulation is important to the overall control of heat loss. To minimize this loss, any exposed steam or hot water piping should be covered. (See Figure A2-1). Any bad insulation around heat-producing equipment should be recovered.

Do office areas have through-the-wall fan coil units? In many cases these will have only a fan speed control. Consider conversion to individual thermostatic control. This will *allow the individual* to govern room temperature and provide better energy control. Air filters on these units should also be checked to ensure:

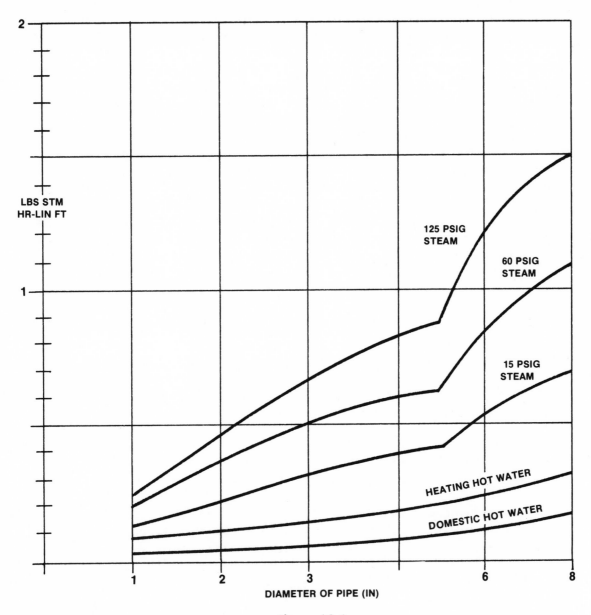

Figure A2-1.

Energy Saved by Insulating Bare Heating Piping with 1″ Thick Fiberglass Insulation.

1.   That they are always kept clean.
2.   That they are the proper type that allows maximum air flow over the coils.

   Over the years many areas are renovated and rearranged, which could result in an unbalanced air distribution system. You may want to have the entire system rebalanced. Though this is an expensive item it may save heating bills by eliminating hot spots and improving no-heat areas. Reheat coils and heat exchanges in ductwork are often overlooked. They should be checked and cleaned for proper operation.

Outside air dampers should be checked for proper closure in the winter. Dampers should be maintained to operate freely and thus prevent unwanted cold air from entering the building.

Roof exhaust fans should also be shut down during building shutdown to reduce emission of conditioned air.

Hot water systems should be closely scrutinized. System temperature settings should be reduced from the typical design setting of 180°F to 120°F, except where required by code. Leaky hot water faucets equate to a loss of 700 gallons per year for every drop per second. Fix those leaks now. Installation of water flow restrictors in the domestic water system will further curtail water waste usage. Insulating hot water piping where exposed will also reduce radiation loss.

V. HEAT RECOVERY TECHNIQUES

Various heat recovery techniques are available to help recapture lost heat:

- Thermal wheels—Figure A2-2
- Waste heat incinerators—Figure A2-7
- Heat of light systems
- Heat pumps—Figure A2-5
- Heat exchangers—Figures A2-3, A2-4, A2-6

Thermal wheels work on the basis of reclaiming conditioned air energy prior to its being exhausted. The thermal wheels act as a rotary heat exchanger, collecting energy from an exhaust duct and transferring it back to the air supply. Thus, less BTUs need be expended to initially condition incoming outside air.

Incinerators may be a source of energy where waste with a high-heating value is available. Incinerators work on the principle of recovering hot combustion gases while they pass through one or more heat exchanges located in the exhaust flue.

Energy used in building lighting systems can be recaptured by returning room air through the lighting fixture. This air runs over the lamp and ballast into ductwork or a ceiling planum, and is then recirculated by the air-handling system.

Heat pumps basically work in a mode opposite of an air conditioning system. Its components are a compressor, condensor, evaporator, and reversing valve. During the winter, heat is extracted from the condenser air and compressor. This heat source is then transfered to the supply air. During the summer, heat is removed from the supply air and rejected by the condenser.

Waste heat from hot liquid steam or gases can be recovered and used to heat up other parts of the facility by strategic use of heat exchangers.

A2-2

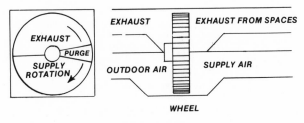

Figure A2-2.

Thermal Wheel.

A2-3

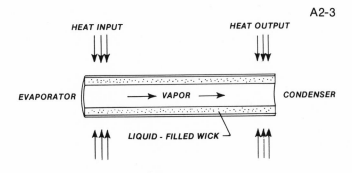

Figure A2-3.

Heat Pipe.

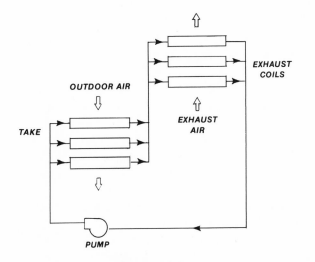

Figure A2-4.

Run-Around Coil System.

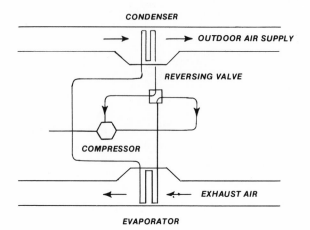

Figure A2-5.

Heat Pump-Heating Mode.

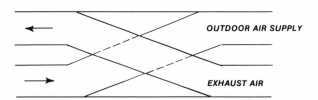

Figure A2-6.

Air-to-Air Heat Exchanger.

VI. KITCHEN OPERATIONS

Paramount to an energy-conserving kitchen operation is the orientation of food service personnel to the proper operation of kitchen equipment. They must also be educated in the area of peak load reduction to electrical systems.

Clean refrigeration condensers, heating coils, and exhaust air filters all contribute to energy reduction.

Some other helpful hints in the area of kitchen energy conservation include:

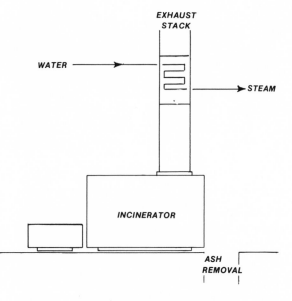

Figure A2-7.

Waste Heat Incinerator.

- Using correct temperatures for each kind of food prepared for oven cooking.
- Placing several foods in the oven at the same time that have compatible temperatures and cooking times.
- Using flat-bottomed pans that fit the cooking element being used will reduce wasted heat.
- Cutting off oven heat a few minutes before the food is cooked. The residual heat will finish the job.
- Keeping freezers full, but not tightly packed. This uses less electricity to maintain desired temperature.
- Keeping cooler doors closed. Check worn door seals and replace where necessary.
- Set refrigeration temperatures at 36°F and freezer temperatures at 0°F because lower temperatures waste power.
- Be sure the dishwasher has a full load before it is turned on. Intermittent usage wastes electricity and water. Scrape dishes, but do not rinse them, before loading them into the washer. Let dishes air-dry.

VII. PEOPLE

People are the key ingredient to all that has been said so far. Without their being educated, without their decisions, without their total cooperation, not one penny will be saved.

Take a look at some energy-saving items that can be used as part of an overall program for employees of your facility. When entering or leaving the building, close the exterior doors promptly to keep the conditioned air inside the building.

During the winter let the sunshine in! Open Venetian blinds and draperies during the day, but be sure to close them at night to help reduce heat loss through the glass. During the summer, draw Venetian blinds and drapes closed to cut down on solar load.

Establish a working liaison with the local utility company to develop programs for energy conservation. They will be more than happy to assist you in your plans. Set up bulletin boards and posters for employees to emphasize energy conservation. Encourage office and operating personnel to participate in the overall program to reduce energy use. Posting charts and bar graphs to indicate monthly energy consumption and demand can provide incentive to join in energy conservation. Report corrective measures taken to stimulate participation in the program.

Educate all building occupants. Solicit ideas from them for specific ways of reducing energy requirements. Appoint employees by work area to be responsible for a continuing conservation effort. Request employees to report energy wasting sources immediately, such as a leak or excessive lighting. Stress that every step taken to conserve energy means that much saved in operating costs, which in turn pays their wages. Formulate an indoctrination program and plan to include all employees in energy awareness.

If possible, establish an incentive program for employees to conserve energy. For example, set up an energy idea award program. Each month someone can be selected as the person giving the best energy-saving idea that is put into effect. In doing this, employees become fully aware of the situation, their support is obtained, and you will eventually realize a reduction in utility bills.

CONCLUSION

Total efficiency in energy conservation is achieved only by careful examination of all factors. A good energy conservation program must consider all of the above items to achieve maximum possible savings.

All energy-saving programs are primarily the responsibility of individual efforts. No firm, regardless of its expertise, can achieve the goals set forth, without the cooperation and support of the organization, its management, and its people.

Index